Proceedings of the US–Europe Workshop on Sensors and Smart Structures Technology

Proceedings of the US–Europe Workshop on Sensors and Smart Structures Technology

Organized by
National Science Foundation (NSF)
Arlington, Virginia, USA

European Science Foundation (ESF)
Strasbourg, France

Edited by

Lucia Faravelli
University of Pavia, Italy

Billie F. Spencer Jr.
University of Illinois, USA

WILEY

Phone (+44) 1243 779777

Email (for orders and customer service enquires): cs-books@wiley.co.uk
Visit our Home Page on www.wiley.co.uk or www.wiley.com

Other Wiley Editorial Offices

John Wiley & Sons, Inc. 111 River Street, Hoboken, NJ 07030, USA

Jossey-Bass, 989 Market Street, San Francisco, CA 94103-1741, USA

Wiley-VCH Verlag GmbH, Pappellaee 3, D-69469 Weinheim, Germany

John Wiley & Sons Australia, Ltd, 33 Park Road, Milton, Queensland, 4064, Australia

John Wiley & Sons (Asia) Pte Ltd, 2 Clementi Loop #02-01, Jin Xing Distripark, Singapore 129809

John Wiley & Sons Canada Ltd, 22 Worcester Road, Etobicoke, Ontario, Canada, M9W 1L1

British Library Cataloguing in Publication Data

A catalogue record for this book is available from the British Library

ISBN 0 471 48980 8

Produced from camera-ready copy supplied by the authors.
Printed and bound in Great Britain by Antony Rowe, Chippenham, Wiltshire.
This book is printed on acid-free paper responsibly manufactured from sustainable forestry in which at least two trees are planted for each one used for paper production.

CONTENTS

Preface..vi
i
Acknowledgements..viii

SECTION 1 1

Flexibility-Based Damage Localization Employing Ambient Vibration
(Spencer and Gao) 3
Active Buffetting Vibration Alleviation - Demonstration of Intelligent Aircraft Structure for Vibration and Dynamic Load Alleviation
(Becker) 9
Wireless Sensor Networks for Structural Health Monitoring and Hazard Mitigation
(Horton) 19
Wireless Communications between Standalone Semiactive Control Devices
(Faravelli and Rossi) 25
R-SHAPE: A Real-Time Structural Health and Performance Evaluation System
(Iwan) 33
Damage Identification for Structures Using Hilbert – Huang Spectral Analysis
(Yang) 39
Fiber Optic Sensors in Civil Structures
(Ansari) 45
Fiber Optic Sensors for Long Term Global Structural Monitoring
(Inaudi and Glisic) 51

SECTION 2 63

Large-Scale Tests on Smart Structures for their Performance Verification
(Magonette) 65
Vibration Suppressionby Energy Pumping
(Bergman) 73
Signal Analisis and Artificial Intelligence in Structural Monitoring and Diagnosis
(De Stefano) 83
Use of Nonparametric Approaches for Structural Health Monitoring of Dampers
(Wolfe) 94
Seismic Response of Elevators in Building and Their Sensor Needs
(Singh and Rildova) 97
Efficacy of Optic fiber Strain Sensor for Determination of Internal Strain in 3-D Braided Composites
(Yuan) 103
Low Frequency PVDF Sensors for Condition Monitoring of Structures
(Lloyd and Wang) 109

Competing Risks in Stochastic Modeling of Structure Deterioration
(Yang) 119

SECTION 3 125

Implementation of New Technology in Inspection and Monitoring
(Yang) 127
Satellite Images for Infrastructure Management
(Casciati) 133
Dynamic Piezoelectric Shape Control Applied to Shells of Revolution with Translatory Support Excitation
(Irschik) 139
Electro-Magneto_Acoustic Transducers for Monitoring and Health Assesment of Metal Structures
(Shouareshi and Lim) 149
Seismic Protection of Historical Masonry Structures by Using Control Techniques
(Syrmakezis) 155
Passivity Based Control of Piezoelectric Structures
(Kugi and Schlacher) 163

WORKING GROUP REPORTS 177
Report ONE 179
Report TWO 181

WORKSHOP RESOLUTION 183

Author Index..........187

PREFACE

In the last few years, significant progress has been made in the area of sensing technology and structural health monitoring/condition assessment in the US and Europe. Innovative concepts involving new hardware, algorithms, and software have been proposed. There have also been several full scale trial implementations of densely sensor instrumented infrastructures and health monitoring systems, as well as case studies on bridges in Europe and in the US. There is much we can learn from each other, particularly in the areas of experimental verification on small, medium, large and full scale projects. Moreover, a common framework for expanded future joint research can be developed on the increased understanding achieved through mutual learning.

The program of the workshop held in Como and Somma Lombardo, Italy, on April 12 and 13, 2002, respectively, consisted of seminar sessions on several themes, including innovative sensing hardware, advances in wireless technology, and damage detection/characterization and condition assessment methodologies. Subsequently, there were several workshop sessions to summarize the status of the sensors and smart structures technologies in these topics, to identify the compelling research issues, and to formulate an action plan with recommendations for development and implementation through possible collaborative research projects and sharing of scientific data. The proceedings, as well as the final report of the workshop, are here prepared and made available to the research and practitioner communities.

Lucia Faravelli and B.F. Spencer, Jr.

Editors

ACKNOWLEDGMENTS

The organizes of the US-European seminar/workshop on the topic of Cooperative US-Europe Research in Sensors and Smart Structures Technology gratefully acknowledge the generous financial support provided by:

US National Science Foundation (NSF);
European Science Foundation (ESF)
Istituto di Studi Superiori dell'Insubria "Gerolamo Cardano"
Fondazione Visconti di San Vito

The morning of Saturday, April 2002, the workshop was held in the magnificent environment of the Visconti Castle in Somma Lombardo, thanks to the foundation Visconti di San Vito and its delegates Cristina Bertacchi and Michela Grisoni.

The Editors acknowledge the cooperation of the Working Group Chairmen and the assistance of the Organizing Committee of the 3rd World Conference on Structural Control, held in Villa Olmo from April 7 to April 11, 2002.

The Editors also thanks the cooperation of Wendy Hunter, of John Wiley & Sons Ltd. and Danilo Miozzari, of the University of Pavia, for helping them in assembling the edited material.

Europe Workshop On Sensors and Smart Structures Technology, Somma Lombardo (VA) April 13, 2002

The magnificent environment of the Visconti Castle owned by Fondazione Visconti di San Vito

Section 1

Flexibility-Based Damage Localization Employing Ambient Vibration

B.F. Spencer, Jr.[1] and Yong Gao[1]

[1]*Department of Civil Engineering and Geological Sciences, University of Notre Dame, Notre Dame, IN 46556, USA*

ABSTRACT

In recent years, Structural Health Monitoring (SHM) has emerged as a new research area in civil engineering. Most existing health monitoring methodologies require measurement of inputs for implementation. However, in many cases, there is no easy way to measure these excitations – or alternatively to excite the structure. Therefore, SHM methods based on ambient vibration become important in civil engineering. In this paper, an approach is proposed to extend the Damage Location Vector (DLV) method to handle the ambient vibration case. Here, the flexibility-matrix based method is combined with a modal expansion technique and an approach to select the analytical model for this modal expansion technique. Finally, a numerical example that analyzes a truss structure with limited sensors and effect of noise is provided to verify the efficacy of the proposed approach.

1. INTRODUCTION

One class of SHM strategies measures the change in frequencies to determine structure damage. Vandiver (1975) examined the change in resonant frequencies due to the damage in local elements. Most recently, Cha and Tuck-lee (2000) examined the change in frequency response data; this information was then used to update structural system parameters. West (1984) was perhaps the first to implement systematic use of mode shape information for localization of structural damage without the use of a prior finite element model. Pandey and Biswas (1994, 1995) presented a damage detection and localization method based on changes in the measured flexibility of the structure. Based on the change of the flexibility matrix, Bernal (2002) computed a set of load vectors, designated as damage location vectors to detect damage.

Although these methods can be effective, most require measurement of the input and at least one co-located sensor actuator pair in the system in order to obtain the required mass-normalized modes (Alvin and Park, 1994). In numerous cases, the use of an impulse hammer or a rotating unbalanced vibrator as an exciter is impossible. Ambient vibration has become an important source of excitation for SHM. Bernal (2001) presented an initial study to extend the Damage Location Vector (DLV) method (Bernal 2002) to handle the ambient vibration case.

In this paper, a new approach that extends the DLV method to tackle the ambient vibration case is proposed. For the case of ambient vibration, no inputs are known so the flexibility matrix cannot be constructed directly using existing approaches. This problem is circumvented via a modal expansion technique with several proposed modifications, allowing the flexibility matrix to be constructed, and then the DLV method to be applied. Appropriate strategies for damage detection are proposed for single and multiple damage scenarios. A numerical example is included to demonstrate the efficacy of the proposed approach.

2. PROBLEM FORMULATION

The basic principle of the DLV method (Bernal 2002) is the computation of a set of vectors, the so-called damage location vectors (DLVs), which have the property of inducing stress fields whose magnitudes are zero in damaged elements. For a linear structure, the flexibility matrices at sensor locations before and after damage are denoted as $\mathbf{F}_u$ and $\mathbf{F}_d$, respectively. If we collect all the linearly-independent load vectors $\mathbf{L}$, which produce same displacements at sensor locations, they satisfy following equations

$$\mathbf{F}_d\mathbf{L} = \mathbf{F}_u\mathbf{L} \qquad \text{or} \qquad (\mathbf{F}_d - \mathbf{F}_u)\mathbf{L} = 0 \tag{1}$$

The singular value decomposition (SVD) of matrix $\mathbf{F}_\Delta = (\mathbf{F}_d - \mathbf{F}_u)$ leads to

$$\mathbf{F}_\Delta = \begin{bmatrix}\mathbf{U}_1 & \mathbf{U}_0\end{bmatrix}\begin{bmatrix}\mathbf{S}_1 & 0\\ 0 & 0\end{bmatrix}\begin{bmatrix}\mathbf{V}_1 & \mathbf{V}_0\end{bmatrix}^T \tag{2}$$

or, equivalently

$$\begin{bmatrix}\mathbf{F}_\Delta\mathbf{V}_1 & \mathbf{F}_\Delta\mathbf{V}_0\end{bmatrix} = \begin{bmatrix}\mathbf{U}_1\mathbf{S} & 0\end{bmatrix} \tag{3}$$

Eqs. (1) and (3) show that $\mathbf{L} = \mathbf{V}_0$, *i.e.*, $\mathbf{L}$ can be obtained from the SVD of matrix $\mathbf{F}_\Delta$. From the definition of DLVs, one can see that DLVs also satisfy Eq. (1); because if DLVs induce no stress in damaged elements, the damage of those elements does not affect the displacements at sensor locations. That is to say, DLVs are indeed the vectors in $\mathbf{L}$. In reality, stresses induced by DLVs in damaged elements may not be exactly zero because of noises and uncertainties. Reasonable threshold should be selected to indicate the damaged element. An index *svn* was proposed by Bernal (2002) to select the DLVs from the SVD of matrix $\mathbf{F}_\Delta$.

To apply the DLV method to locate the damage, the flexibility matrix needs to be constructed. Because the measured DOFs are smaller than the DOFs of the structure, and not all modes are identified, once modal parameters are obtained, a modal expansion technique is needed to construct the flexibility matrix.

The expansion approach suggested by Lipkins and Vandeurzen (1987) is applied herein. When same number of experimental mode shapes and analytical mode shapes are used, the equations can be written as

$$\begin{bmatrix} [\varphi_1^E]_{n \times p} \\ [\varphi_2^E]_{(N-n) \times p} \end{bmatrix} = \begin{bmatrix} [\varphi_1^A]_{n \times p} \\ [\varphi_2^A]_{(N-n) \times p} \end{bmatrix} \lambda_{p \times p} \tag{4}$$

in which, φ^A are the analytical mode shapes, and φ^E are the experimental mode shapes; N = number of DOFs of the structure; n = number of measured DOFs, *i.e.*, number of sensors; p = number of identified mode shapes; and λ = coefficient matrix. As long as $n \geq p$, the coefficient matrix λ can be obtained from Eq. (4) in a least-squares sense as

$$\lambda = ([\varphi_1^A]^T [\varphi_1^A])^{-1} [\varphi_1^A]^T [\varphi_1^E] \tag{5}$$

Mode shapes at unmeasured DOFs (*i.e.*, $[\varphi_2^E]$) can then easily be computed from Eq. (4). Then, the flexibility matrix can be constructed using

$$\mathbf{F}_1 = (\varphi_1 \mathbf{v}^{-1}) \omega^{-2} (\varphi_1 \mathbf{v}^{-1})^T, \ \mathbf{v} = (\varphi^T \mathbf{M} \varphi)^{1/2} \tag{6}$$

From Eq. (4), the experimental mode shapes are presumed as a linear combination of analytical mode shapes. One way to select these analytical mode shapes for a given analytical model is based on the Modal Assurance Criterion (MAC), suggested by Ewins (1985), which is defined as

$$\mathrm{MAC}(\{\phi^E\}_i, \{\phi^A\}_j) = \frac{\left|\{\phi^E\}_i^T \{\phi^A\}_j\right|^2}{(\{\phi^E\}_i^T \{\phi^E\}_i)(\{\phi^A\}_j^T \{\phi^A\}_j)} \tag{7}$$

Analytical mode shapes with a MAC value closed to 1.0 are selected. Then, the Total Modal Assurance Criterion (TMAC) is used to determine the analytical model, which give the analytical mode shapes in Eq. (4)

$$\mathrm{TMAC} = \prod_{i=1}^{p} (\mathrm{MAC}(\{\phi^E\}_i, \{\phi^A\}_i)) \tag{8}$$

in which p = number of identified modes.

The key question then lies in determining exactly how to employ the TMAC to select the analytical model, denoted as the DAM – the damaged analytical model. First, the undamaged analytical model is constructed. Then the undamaged analytical model with one element damaged will be selected as the DAM. There are two steps to follow in choosing the DAM. The first step attempts to determine which element, if damaged, may generate the DAM. In this step, each element is damaged to a few different damage extents (for example, 8 evenly-distributed damage extents for each element). Results are compared, and the analytical model (*e.g.* the ith element is damaged) with the highest TMAC value is chosen. Then, in the second step, a wide range of damage extents (*e.g.* 100 damage extents) is evaluated for the ith element alone to choose the damage extent corresponding to the highest TMAC. After these two steps, *i.e.,* selecting the damage element and then the damage extent, the DAM is determined. Mode shapes at unmeasured DOFs then can be calculated from Eq. (4). The flexibility matrix at sensor locations can be obtained easily from Eq. (6). After pre- and post-damage flexibility matrices are obtained, the DLV method can be applied to locate the damaged element.

Note, however, that if the damaged element indicated by the DLV method does not match the damage element in the DAM, it is necessary to reselect the DAM corresponding to the second-highest TMAC value, third-highest, etc. This step is important, as the DAM that carries the correct damaged element does not necessarily produce the highest TMAC value because of noise and uncertainties. Additionally, TMAC values reflect only changes in mode shapes, while the DLV method includes both frequency and mode shape information through the flexibility matrix. Also, assuming the analytical model is reasonably accurate, if the DAM includes the correct damaged element, the DLV method will indicate the correct damaged element. Again, high confidence is obtained when the damaged element in the DAM is same as that indicated by the DLV method.

At this point, the proposed approach identifies only one damaged element in the structure. For the multiple damage case, iterative searching is necessary. First, one of damaged elements is identified based on the approach outlined above. Then the identified DAM is denoted as the new baseline model (undamaged analytical model), and the above approach is repeated to detect the next damaged element. The remaining problem lies in determining when to stop the iterative search. An idea proposed here is based on the fact that the DAM with the same number of damaged elements as the damaged structure should have more information than others. An index, so-called Averaged Highest TMAC value (AHTMAC) is proposed here as the flag

$$\mathrm{AHTMAC} = \left(\sum_{j=1}^{N_{\mathrm{E}}} \mathrm{HTMAC}_j \right) / \; N_{\mathrm{E}} \tag{9}$$

in which HTMAC_j = the highest TMAC value observed when the jth element of DAM is damaged; and N_{E} = total number of elements in the DAM. If the index AHTMAC for a new search is smaller than the previous one, iteration should stop.

3. NUMERICAL EXAMPLE

The proposed detection approach is demonstrated using a planar truss structure, as shown in Fig. 1. A similar truss was considered by Bernal (2002).

In this numerical example, limited sensors are employed (9 sensors compared with 40 DOFs), and a 5% RMS noise is added to the outputs. Elements are connected at pinned joints, each having two DOFs. To simulate the ambient vibration case, inputs are not collected. Two categories of damages are considered:

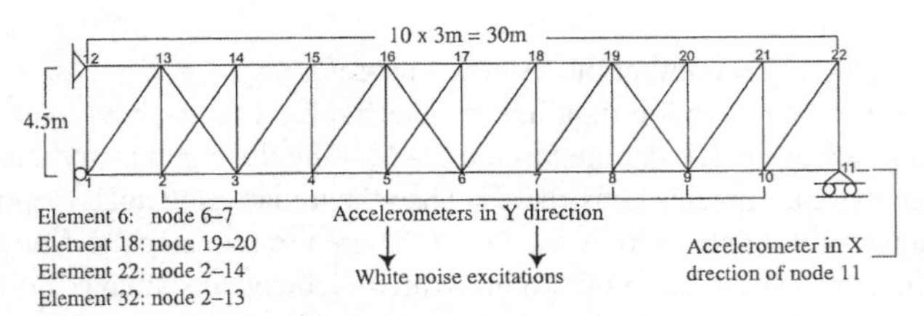

Figure 1. 44-bar planar truss

- *Case 1:* Single damage scenario – 15% stiffness reduction in a single member.
- *Case 2:* Multiple damage scenario – 30% stiffness reduction in elements 18 and 22.

First, modal parameters are obtained from measured data. TMAC is then applied to find the DAM for modal expansion. Due to limited space, results for only two cases in the single damage scenario are shown herein: 1) only element 6 is damaged; and 2) only element 32 is damaged. Fig. 2 provides stresses induced by the DLVs, which clearly show element 6 is damaged. In this case, the DAM corresponding to the highest TMAC value has same damage element as the damaged structure.

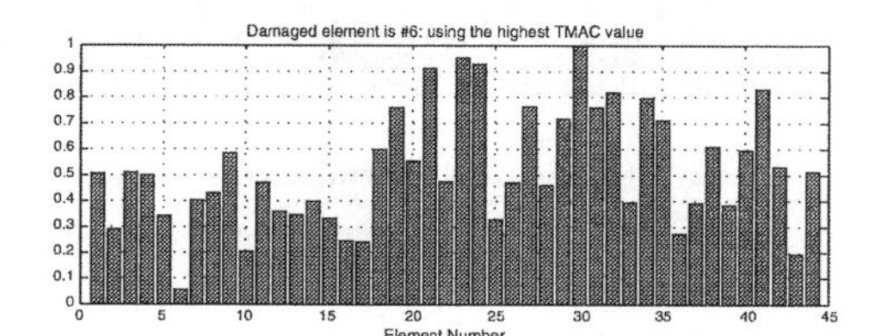

Figure 2. Normalized accumulative stress (case 1: element 6)

The case with damage in element 32 represents a more general situation; that is, the DAM corresponding to the highest TMAC value does not have the same element as the damaged structure. However, after repeating selection of the DAM, the damaged element in the structure, in the DAM, and indicated by the DLV method coincide. Results for the case with damage in element 32 are shown in Fig. 3. The left graph in Fig. 3 shows results with the DAM corresponding to the highest TMAC value. The DAM has a 6% stiffness reduction in element 17. We see that all normalized accumulative stresses are relatively large, and there is no clear information about which element is damaged. The right graph in Fig. 3 displays results after repeating the procedure to select the DAM. In this case, the DAM indicates a 19% stiffness reduction in element 32. Fig. 3 shows that element 32 has a relative small accumulative stress. The damaged element is thus predicted correctly.

For multiple damage scenario, the case when elements 18 and 22 both have a 30% stiffness reduction is presented here. As discussed previously, for multiple damage scenario, in each step, one damaged element is determined. First, element 18 is identified as shown in the left graph of Fig. 4. After element 18 is identified, the model with element 18 damaged is designated as the new baseline model. Then, this baseline model is used to detect the damage in element 22. Results for element 22 are shown in the right graph of Fig. 4, in which the normalized accumulative stress of element 22 is reasonably small. Therefore, the multiple damage case can also be handled by the proposed approach.

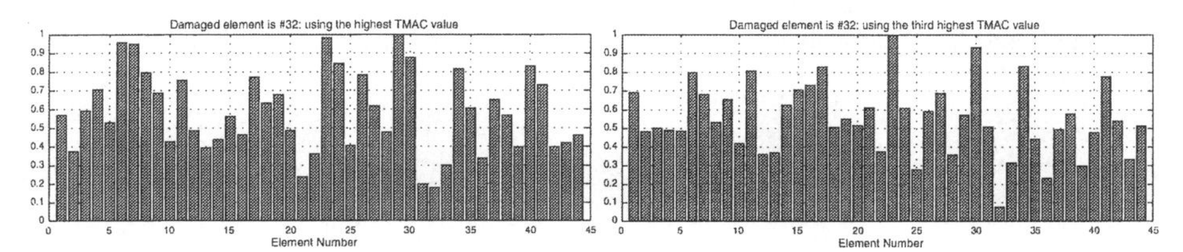

Figure 3. Normalized accumulative stress (case 1: element 32)

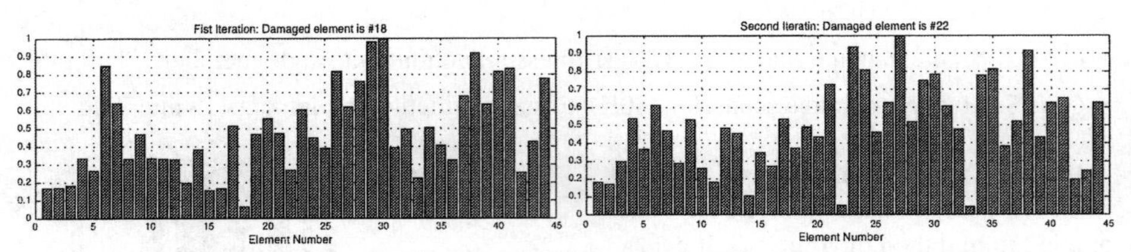

Figure 4. Normalized accumulative stress (case 2: element 18 and 22)

4. CONCLUSIONS

An approach that extends the DLV method to include the ambient vibration case is presented in this paper. By applying the modal expansion method with the proposed approach for selecting the analytical model, the flexibility matrix can be constructed. The DLV method can then be applied to locate the damage in the structure. The numerical example produces results that are quite reasonable for damage that causes as small as a 15% reduction in local element stiffness. The approach works well for both single and multiple damage cases.

REFERENCES

Alvin, K.F. and Park, K.C. (1994) Second-order Structural Identification Procedure via State-Space-Based System Identification. *AIAA Journal*, 32(2): 397–406.

Bernal, D. (2001) Damage Localization in Output-only Systems. *3rd International Workshop in Structural Health Monitoring*, Stanford, California, USA, September.

Bernal, D. (2002) Load Vectors for Damage Localization. *Journal of Engineering Mechanics*, 128(1): 7–14.

Cha, P.D. and Tuck-Lee, J.P. (2000) Updating Structural System Parameters Using Frequency Response Data. *Journal of Engineering Mechanics*, 126(12): 1240–1246.

Ewins, D.J. (1985) *Modal Testing: Theory and Practice*, John Wiley, New York, USA.

Lipkins, J. and Vandeurzen, U. (1987) The Use of Smoothing Techniques for Structural Modification Applications. *Proceedings of 12 International Seminar on Modal Analysis*, S1–3.

Pandey, A.K. and Biswas, M. (1994) Damage Detection in Structures Using Changes in Flexibility. *Journal of Sound and Vibration*, 169(1): 3–17.

Pandey, A.K. and Biswas, M. (1995) Damage Diagnosis of Truss Structures by Estimation of Flexibility Change. *The International Journal of Analytical and Experimental Modal Analysis*, 10(2): 104–117.

Vandiver, J.K. (1975) Detection of Structural Failure on Fixed Platforms by Measurement of Dynamic Response. *Proceedings of the 7th Annual Offshore Technology Conference*, 243–252.

West, W.M. (1984) Illustration of the Use of Modal Assurance Criterion to Detect Structural Changes in an Orbiter Test Specimen. *Proceedings of the Air Force Conference on Aircraft Structural Integrity*, 1–6.

Active Buffeting Vibration Alleviation Demonstration of Intelligent Aircraft Structure for vibration & dynamic load alleviation

Jürgen Becker

EADS –Deutschland GmbH Military Aircraft 81663 Munich, Germany

ABSTRACT

The various fields of the application of intelligent active structures for military aircraft are discussed and results from a special detailed investigation of active structure are presented. The detailed investigation deals with heavy structural vibrations which can be observed at wing and vertical tail of high performance aircraft flying at high angles of attack by vortices originating from wing leading edge and from front fuselage/canard. The resulting wing/fin dynamic loads may lead to increased material fatigue. A number of different passive and active concepts have been investigated to minimize the excitation of the wing/fin or to alleviate the resulting structural vibrations. Active system concepts were suggested as an efficient way for active buffet load alleviation. A collaborative research project was initiated between EADS – D Military Aircraft , the former DaimlerChrysler Aerospace - Military Aircraft Division, the German Aerospace Centre (DLR) and DaimlerChrysler Research and Technology within the framework of the Advanced Aircraft Structures Research Program. Four concepts were investigated in detail within this project: An active rudder, an active auxiliary rudder, a piezo-controlled interface and a system of surface-mounted or structurally integrated piezoelectric patch actuators.
The feasibility of all these concepts could be proven and their performance could be assessed in an extensive theoretical analysis that involved the complete aircraft system, as well as in wind tunnel tests on the rudder concepts and, for the piezo-controlled concepts, in tests on a laboratory demonstrator that was conceived, designed and manufactured to be dynamically equivalent to a typical fighter fin. In addition, a materials qualification program was initiated in order to demonstrate the compatibility of structures with integrated piezo-ceramic actuators with the requirements for fighter aircraft structure. In this way the maturity of this new technology could be shown.

1. INTRODUCTION

Several investigations of active structure application for military aircraft have been performed in the past by EADS Military Aircraft. The results of these studies have been documented and reported in the past, (ref. 1-8).
The general purpose of Active Structure is to mainly found in the fields of vibration alleviation, dynamic load reduction, acoustic noise reduction and fatigue loads reduction. Improvement of performance can be achieved by shape control through drag reduction. Enhanced aircraft stabilisation and maneouvrebility might be possible. In addition health monitoring systems shall be feasible using intelligent structure.

EADS Deutschland Military Aircraft Division currently has performed research work in the field of advanced aircraft structures in a joint research program with Daimler Benz Research (DB - F4T) and the DLR Institut für Strukturmechanik , Braunschweig. Especially concepts for vibration control have been investigated in the development phase of the technology program of adaptive structures. Aerodynamic concepts for vibration alleviation, a rudder and an auxiliar rudder concept has been investigated by EADS Military Aircraft , an integrated piezo concept was investigated together with EADS Research and a piezo-interface concept was investigated in collaboration with the DLR Braunschweig. The task was performed in a pre-development phase , which contained the set up of specifications and description of design concepts and tools, and a main phase consisting of validation of concepts by tests, analytical modeling updates and controller design and validation. The adaptive vibration control systems were aimed to reduce the vibration levels induced by gust and by buffet at high incidence on wing , fuselage and fin of military aircraft. Aim of the systems is decrease of fatigue loads or extension of flight envelope with respect to incidence.

The development of vibration alleviation control systems for a modern military aircraft is strongly influenced by flight mechanic, flight control and aero- servo- elastic effects. The flexible aircraft behavior has significant effects on the active vibration alleviation control system. The sensor signals i.e. accelerometer signals on wing, fuselage and fin and the signals of the Aircraft Motion Sensor Unit - the gyro platform - contain besides the necessary information of rigid aircraft rates and accelerations the flexible aircraft rates and accelerations in the frequencies of the aircraft elastic modes. The 'flexible' accelerations measured by the accelerometers are passed through the active vibration flight control system control paths, they are multiplied by appropriate gains filters and inserted in the control surface actuator (rudder or auxiliary rudder) or integrated piezo, interface actuator inputs. The flexible aircraft is excited by the high frequency actuation inputs and might therefore experience eliminate aero-servo-elastic instabilities i.e. flutter or limit cycle oscillations, and/or decrease dynamic load and fatigue loads. The adaptive vibration alleviation system design therefore has to minimize all structural coupling effects to avoid aero-elastic instabilities through the available means like optimum sensor positioning, notch filtering. Besides the vibration alleviation also concepts of deformable wing areas had been studied. This paper describes the major aspects, i.e. the field of applicability , problem areas to be considered in the adaptive system for vibration alleviation design with respect to hydraulic and piezoelectric actuator design and total aircraft aero-servo-elastic effects and describes also aspects of clearance procedures for the flight clearance of intelligent structure.

2. DESIGN AND CLEARANCE OF ACTIVE STRUCTURE

A qualification program plan (QPP) was established to define the methodology for the qualification of the active buffet load alleviation system. It detailed the methods required for the qualification procedure.
The development of the active vibration control (AVS) system is performed in several phases with the objective of qualification for flight testing. Phase 1 comprised the technology qualification stage ending with a technology development qualification:

- Definition of a framework of fundamental parameters for the AVS system
- Experimental and/or analytical demonstration of AVS functionality
- Experimental and/or analytical investigation of critical specifications and the compliance of the AVS system with the QPP requirements.

For the active fin buffet alleviation system the QPP identified the following activities that were performed in Phase 1:

- Theoretical investigation of the performance of each of the four concepts using a whole aircraft dynamic model.
- Scaled-model wind-tunnel tests to prove the aerodynamic efficiency of the auxiliary rudder concept.
- Laboratory demonstrator tests on a full scale fin box model to prove the actuating authority of the active interface and the structurally integrated piezoelectric actuator concept.
- A materials qualification program intended to demonstrate the compatibility of the AVS system based on distributed surface-bonded or integrated piezoelectric patches as well as discrete piezoelectric stacks with the requirements and specifications for aircraft applications.

The results of these efforts are compiled in the following section.

3. DESIGN OF ACTIVE FLIGHT CONTROL SYSTEM FOR VIBRATION ALLEVIATION

Design philosophy

The design shall include the derivation of adaptive vibration alleviation (AVS) gains, phase advance filters and notch filters to minimize vibrations and dynamic loads on fin, wing and fuselage structure. In addition the structural coupling effects on total aircraft vibration modes shall be minimized in the AVS optimization process. The AVS shall be designed to cover the full rigid, flexible aircraft frequency range with respect to aircraft rigid mode and structural mode induced dynamic loads and coupling stability requirements on ground and in flight. The structural coupling influences shall be minimized by AVS and FCS notch filters and notch filters in the wing, fin accelerometer signal feedback signals . The AVS shall be designed to be as robust as possible with respect to all possible aircraft configurations and configuration changes. That includes that all structural changes with configuration should be covered by a constant set of filters to avoid system complexity due to configuration switches

for different sets of filters. In addition scheduling of AVS gains and filters with flight conditions should be introduced to cover the flight envelope. In order to avoid problems in the AVS design due to non-linear unsteady elastic mode and control surface aerodynamics and nonlinear piezo-actuator dynamics the elastic mode stability requirements should mainly be based on gain stabilization of the flexible modes. Phase stabilization shall be applied if it can be demonstrated by test.
The AVS design is based upon an analytical model of the total aircraft structure including a linear AVS and FCS model. The analytical model must however be verified through ground test results both from ground resonance and structural coupling testing on aircraft or aircraft components, for example on a component with integrated piezo's or with piezo interface .If an AVS demonstrator is envisaged, verification of model technique has to be performed by in flight flutter and structural coupling testing. The model should be updated by the test results for different configurations.

3.1 Design Requirements

Stability Requirements

The design requirements are primarily stability requirements for all adaptive vibration modes from AVS and flight control rigid/flexible aircraft modes. The stability is achieved by the introduction of notch filters.. The Military Specification MIL-F-9490 D for FCS requirements shall be met.

Vibration Alleviation and Dynamic Loads Requirements

The aircraft vibrations on all locations shall meet the requirements of the aircraft without AVS. The dynamic loads with AVS shall meet the allowable load for the aircraft without AVS for the purpose of increasing the existing incidence range. The dynamic loads with AVS shall reduce the dynamic loads for the aircraft with AVS for the purpose of increasing the aircraft life cycle.

Flutter Requirements

The FCS design for AVS has to fulfill the flutter requirements of the aircraft without AVS and FCS. The aircraft with AVS and FCS shall meet the 15 % flutter speed margin as well as the minimum elastic mode damping requirements as described in Military Specification MIL-A-8870 B.

Active Structure Design and Clearance requirements

They are described in Ref.1, 6 showing the general specification , the coupon test requirements, testbox test requirements.

3.2 Design Tools for AVS

The AVS design for the flexible aircraft is possible with the assumption that the aircraft characteristics are predictable to the necessary accuracy to optimize filters which meet the requirements, details of dynamic analytical modeling are described in Ref. 1, 2.

4. DEMONSTRATION OF QUALIFICATION

An extensive testing demonstration and analysis program was conducted in order to complete the technology qualification stage (phase I) and also a first preliminary qualification stage (phase II) demonstrating a principle compliance of the systems with the specifications that are required. A formal qualification (phase III) was not performed and will only have to be conducted an actual flight demonstration program. However, based on the status achieved in the preliminary qualification valuable conclusions can be drawn with respect to the tasks that still have to be completed for a formal qualification. The achieved degree of qualification will be summarized in brief in the following sections.

4.1 Distribuited Piezoelectric and piezo-interface actuator concept

As the use of piezoceramic actuators involved the introduction of a new composite material into the aircraft that has so far not been extensively qualified and certified for use in aircraft applications a detailed material qualification program was conceived in order to

- determine material properties of piezoceramic actuators from material samples
- experimentally characterize actuators and smart composites to provide input data for model calculations.
- perform tests on composite test coupons to demonstrate the conformity of intelligent materials systems with specifications imposed upon them for their integration in aircraft.
- identify, define and perform non-classical tests on smart composite systems in order to test and assure the complete functionality of the intelligent material system.
- develop and demonstrate concepts to integrate smart materials and systems into aircraft design, manufacture and maintenance procedures and processes.

The standardized actuators used in the coupon tests as well as the customized actuator modules on the fin box demonstrator were manufactured by Active Control eXperts (ACX) using standard PZT-5A ceramic material.
The following qualification was achieved using coupon tests and piezo-stack actuator tests: the results of the qualification achieved for integrated piezo - as well as for the piezo - interface and the wind tunnel test for the auxiliary rudder concept are summarised in ref. 5, 6. Both concepts have reached a sufficient stage of maturity for a subsequent demonstrator program.

The following qualification was achieved using fin box tests with integrated piezos and piezo-stack actuators:

Fin box demonstrator test phase 1 – System identification of the fin box without piezoelectric actuators.
Update of fin box FE model based on test phase 1. The original FE model showed good agreement with the experiment, an update had only to be conducted with respect to the local CFC panel thickness.
Fin box demonstrator test phase 2 – Excitation of structural vibrations through piezoelectric actuators and piezoelectric stack actuators. The predicted response levels could be shown.
Update of fin box FEM model based on test phase 2 turned out to be not necessary as the comparison between model predictions and experimental results showed excellent agreement.
Fin box demonstrator test phase 3/4 – Closed loop damping of the first bending and first torsion mode. The load alleviation due to a simulated buffet excitation compared well with the values predicted by the model.

4.2 Rudder and Auxiliary Rudder Concept

As the rudder and auxiliary rudder concept are based on materials, structures and concepts that are well established in aircraft only analyses and system level tests based on the validated analytical aircraft model had to be performed. For the rudder itself results from ground and flight tests could be used for model update. For the auxiliary rudder concept the aerodynamic forces had to be determined from wind-tunnel measurements at the Technical University of Munich [ref 7, 8].
The following analyses and system level tests were conducted for the rudder concept:
A **complete aircraft model** with the rudder was established. Control laws for the rudder concept were determined and power requirements for the buffet alleviation system were documented.
The following analyses and system level tests were conducted for the auxiliary rudder concept:
Establishment of a **finite element model** for the wind-tunnel model with auxiliary rudder.
Wind-tunnel tests for the auxiliary rudder concept phase 1 – Excitation through the auxiliary rudder for angles of attack of up to 31 degrees. These experiments conclusively demonstrated the assumptions about the aerodynamic phenomena that had been used in the control law design as well as the effectiveness of the auxiliary rudder at least up to an angle of attack of 31 degrees.
Update of the fin FE model based on test phase 1.
Wind-tunnel tests for the auxiliary rudder concept phase 2 – Excitation of the fin through buffeting, auxiliary rudder for vibration damping (closed loop testing). A reduction of the fin-tip acceleration caused by buffeting of 60 % could be shown for the closed loop control for all angles of attack up to 31 degrees.

5. RESULTS FOR TOTAL AIRCRAFT

Analytical investigations have been performed for the different AVS concepts using the analytical model of a total modern fighter aircraft including a description of AVS and FCS. Open loop frequency response calculations have been calculated first for the different systems with normalized AVS gains and filters as design input for the AVS gain and filter optimization. A first preliminary AVS control law optimization was performed for theaerodynamic systems (rudder and auxiliary rudder concept) and AVS closed loop calculations have been carried out after definition of AVS gains and filters in order to demonstrate the effects of AVS in comparison to the aircraft without AVS. For the integrated piezo system and the piezo interface only the excitation of the aircraft via piezo actuators and with piezo interface was investigated. Since all investigation of the different systems were based on the same analytical model , the results achieved could directly be compared. The inputs for generalized integrated piezo's and the generalized piezo interface forces which were used for the excitation of the total aircraft have been generated by EADS and by DLR Braunschweig for the piezo interface. The generalized piezo forces were derived from analytical model calculation of the fin component which includes a FEM representation of the integrated piezo actuators.

Comparison of the different systems

A comparison of the excitation of fin response due to excitation for the different systems is shown in Table 1. The comparison shows that the aerodynamic AVS systems as well as the integrated piezo system show similar excitation levels in the frequency range considered. The comparison is performed on the basis of maximum actuator amplitude and max. Volt input at the same level of applied energy.

6. VALIDATION OF THE TOTAL AIRCRAFT MODEL

The analytical model for the present investigation is validated by on ground test and flight test results. The validation was based on comparisons of predicted and measured on ground GRT and structural coupling test results on ground and in flight. Further validation is needed for the auxiliary rudder concept, this will be performed on a total aircraft windtunnel model. Further validation is also needed for the modeling of integrated piezo concept and of the interface piezo concept. This validation will be performed in a first step on a already existing test box which simulates the aircraft fin structure.

7. CONCLUSION

Within the "Advanced Aircraft Structures" research program four concepts for active buffet-load alleviation were investigated in detail. Benefits and drawbacks of the implementation of the individual concepts were assessed. A preliminary system qualification was performed for all these concepts. Investigations have been performed to detail the achievable alleviation of the systems in relation to maximum buffet induced vibrations for different flight conditions The present total flexible aircraft investigations performed for the different adaptive vibration alleviation system, the rudder and auxiliary rudder concept, the integrated surface piezo and the piezo-interface concept came to the following results.

The analytically investigated different adaptive vibration alleviation systems, rudder, auxiliary rudder and integrated piezo concept show similar high vibration alleviation at the aircraft fin related to equivalent power of the systems.

The results presented for integrated surface piezo and piezo interface concept as well as the auxiliary rudder concept have been verified through test box tests and wind tunnel test in case of auxiliary rudder.

Special tests has been performed to investigate effects of military aircraft environmental conditions on functionality and fatigue aspects of integrated surface piezo's and piezo interface actuation systems.

Nonlinear effects of integrated surface piezo's and piezo interface actuation systems under maximum static and dynamic loading and electric power conditions have been investigated through tests in order to update analytical models for AVS system design.

The aerodynamic systems are at present more realistic than the piezo systems. Main actual problem area of the piezo systems are the volume and the absolutely safe functionality of the power supply systems.

7. REFERENCES

[1] J. Becker, W. Luber
Comparison of piezoelectric systems and aerodynamic systems fort aircraft vibration alleviation-
5th Annual International Symposium on Smart Structures and Materials , March 98, San Diego
[2] J. Simpson
Industrial approach to piezoelectric damping of large fighter aircraft components- 5th Annual International Symposium on Smart
Structures and Materials , March 98, San Diego
[3] Becker J., Schröder W., Dittrich K., Bauer E.,J., Zippold H.
The Advanced Aircraft Structures Program – An Overview6th Annual International Symposium on Smart Structures and Materials , March 99, Newport Beach, California USA
[4] Manser R., Simpson J., Becker J.,Dürr J., Flöth E., Herold-Schmidt U., Stark H., Zaglauer H.W.Finbuffet
alleviation via distributed piezoelectric actuators – full – scale demonstrator tests The Advanced Aircraft Structures Program – An Overview6th Annual International Symposium on Smart Structures and Materials , March 99, Newport Beach, California USA
[5] Becker J., Dittrich K., Manser R., Simpson J. Dürr J., Flöth E., Herold-Schmidt U., Ihler E., Zaglauer H.W.,
Fin-buffet alleviation via distributed piezoelectric actuators: materials qualification program and full scale demonstrator testsAdaptronik-Congress 99, March 99, Postdam, Germany
[6] J.K. Dürr, U. Herold-Schmidt, H. W. Zaglauer, and J. Becker
Active Fin-Buffeting Alleviation for Fighter Aircraft RTO 2000 Braunschweig
[7] Breitsamter, Ch.:
"Aerodynamic Active Vibration Control for Single-Fin Buffeting Alleviation", Deutscher Luft- und Raumfahrtkongress / DGLR Jahrestagung, Berlin, 27-30. Sept. 1999.
[8] Breitsamter, C.; Laschka, B.:

"Aerodynamic Active Control for EF-2000 Fin Buffet Load Alleviation", AIAA 2000-0656, 38th Aerospace Sciences Meeting Exhibit, Reno, NV, 10-13 January 2000.

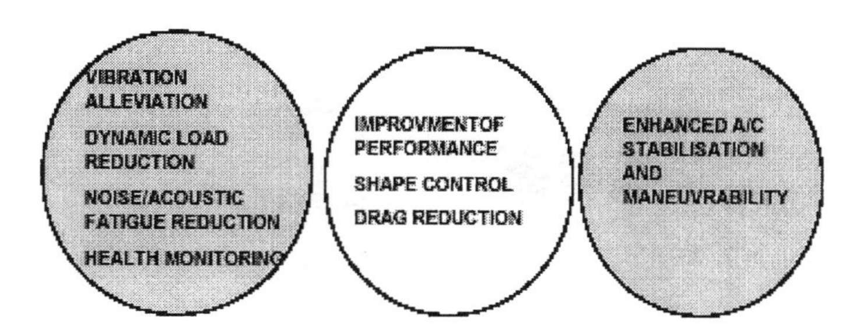

Purpose of Intelligent Aircraft Structure

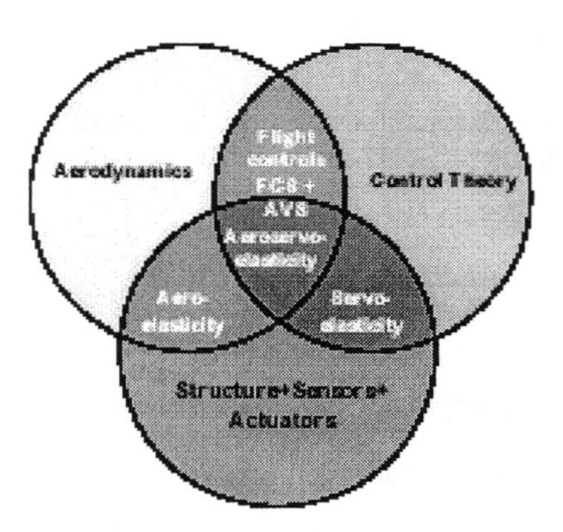

Design of Intelligent Aircraft Structure – Consideration of Interaction of Sensor/ Actuation / Control/ Structure &Aerodynamics

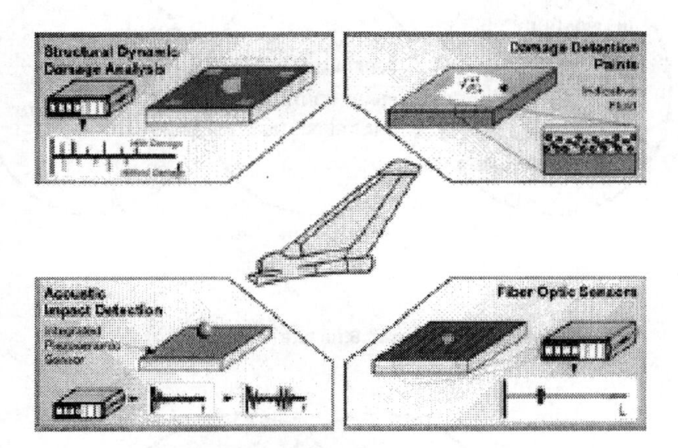

Demonstration of Intelligent Aircraft Structure – for vibration & dynamic load alleviation& health monitoring

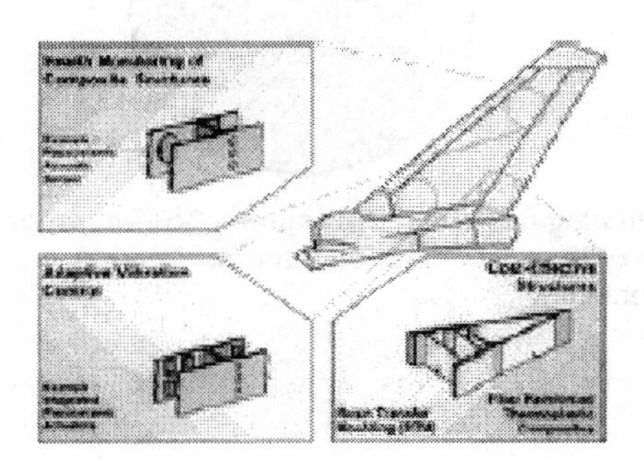

Demonstration of Intelligent Aircraft Structure – for vibration & dynamic load alleviation& health monitoring

Wireless Sensor Networks for Structural Health Monitoring and Hazard Mitigation

[1]Mike A. Horton, [2]Prof. Steve Glaser, [2]Prof. Nick Sitar

[1]*Crossbow Technology*
[2]*University of California at Berkeley*

BACKGROUND:

Imagine a future where buildings and structures contain thousands of smart sensors capable of measuring the building's natural deterioration as well as the response to external events, such as seismic or man-made events providing a continuous view into the structure's health.

We are working to make this vision a reality by developing commercially available, user-extendable wireless sensor products and technologies. These technologies allow for the dense instrumentation of structures at relatively low-cost.

In addition to using silicon and MEMS technology which make the sensors smaller and cheaper, a common software and communications framework is required to allow the sensors to communicate, share information, and estimate the structural state. To facilitate these processes an operating system was developed by UC Berkeley that runs on each sensor node. This operating system is known as TinyOS. TinyOS is open source, which means that the community can add code to the operating system. There are over 100 research groups actively using the TinyOS operating system in all fields of engineering.

Crossbow is working to commercialize these technologies as a series of products that will enable researchers in the structural community to instrument structures at a density level heretofore unseen.

Crossbow has taken the first steps to commercialization by participating in a series of seismic research tests conducted world wide.

TECHNOLOGY:

Before starting the design of a new instrumentation system for structural monitoring, we asked the questions – what are the ideal qualities of a sensor network for structural monitoring ?

These characteristics are:

- Intelligent with Processor
- Reliable
- Capable of High-Fidelity Measurement
- Asynchronous
- Adaptive
- Wireless
- Reprogrammable
- Small
- Low-Cost

The first steps taken to create such a network was building a commercial of the shelf hardware device, that attempts to satisfy these goals. This device is known as a MOTE. The MOTES were created by Professor Kris Pister at UC Berkeley, and commercialized by Crossbow Technology. Subsequently Professor David Culler at UC Berkeley has put together a large effort to design software for MOTES and extend the TinyOS capabilities. The MOTE designs hase gone thru three design generations each time improving the capability

WeC ('we see') – 10Kb/s radio link, 4K flash program space, 128bytes of ram, built in light temperature sensor board.

Renee – 20kb/s radio link, 8K flash program space, 256 bytes of ram, 51-pin connector for external sensors including analog and digital sensors (light, temp, acceleration, user extendable area)

MICA – 40kb/s radio link, 128K flash program, 256 bytes of ram, 64-bit digital ID, 4Mbit data logging storage, 51-pin connector for external sensor including analog and digital sensors (examples - light, temp, acceleration, acoustic, magnetic, GPS, and user-extendable area)

The MICA mote is available from Crossbow as a developer's kit. In addition, Crossbow has begun work on a product custom designed for structural applications known as the CN4000. The CN4000 combines the MICA mote technology with its MEMS based Accelerometer technology. In addition the CN4000 provides a longer range radio and external analog input sensor channels for other variables for example, strain.

All of the wireless sensor products are powered by standard 3V battery technology. In the standard configuration this power is two double AA batteries. The battery life is a function of the usage scenario. All of these sensors have two primary modes of operation Full-on (Pon) and sleep (Poff). The battery life is a function of the percentage of time spent in the Eoff state and the battery capacity.

$$\text{Battery Life} = \text{Energy Capacity} / (\text{Pon}*\text{Ton} + \text{Poff}*\text{Toff})$$

The following table offers some basic statistics for the Renee mote using two standard alkaline AA batteries:
Energy Capacity = 5500 mW * Hours
Pon = 40mW
Poff = 20uW

The battery life is easily computed for various usage scenarios.

100% on -> 5.7 days
1% on -> 545 days

Low-duty cycle operation clearly provides tremendous improvement in battery life.

SENSOR TECHNOLOGY

There is significant research taking place today that uses ambient vibration to predict damage and structural health. Ambient vibration is an attractive quantity to use because it is relatively easy to observe, easy to instrument, and provides adequate data to predict structural damage.

Our sensor network strategy has been to accept all kinds of sensors, focusing on providing the instrumentation and data link back end. However, due to the importance of accelerometer and our core-competence in accelerometers we will discuss their design and nature.

Several characteristics define accelerometer response for seismic research. These quantities are:

Measurement Range
Frequency Range
Noise Floor
Linearity
Temperature Sensitivity
Power Consumption*

For use in long-term sensor networks, with battery power, power consumption is also a crucial parameter.
Crossbow has developed three acceleration modules for seismic research. The three modules achieve different performance capabilities.

	LP Series	**TG Series**	**SM Series**
Measurement Range	2 G	2 G	2 G
Frequency Range	DC-100Hz	DC-200Hz	DC-100Hz
Noise Floor	0.01 G	10e-5 G 10e-6 G	
Linearity	< 0.1%	< 1.0%	< 0.5%
Temperature Sensitivity	< 0.1%	< 1%	< 0.5%
Power Consumption	< 3mA/axis	< 1mA/axis	< 10mA/axis

The LP series uses a capacitive silicon MEMS sensor, built with silicon surface micromachining techniques. It's primary benefit is for general purpose measurements. Because of the surface micromachining technology, the LP has an extremely small die and seismic mass. The resulting cost to produce the device is low. The small mass and tightly coupled leading to a small change in sensitivity over vibration. However, the noise floor of 0.01G is high and in some cases inadequate to capture ambient vibration response. The LP is our best value accelerometer.

The TG Series also uses a capacitive silicon MEMS sensor, built with silicon bulk micromaching techniques. Bulk micromachining leads to a larger seismic mass. The larger seismic mass offers more signal to inertial movement and hence produces a noise floor that is significantly lower than that of the LP Series. The TG Series is often a good compromise between performance and cost. In addition since many data acquisition systems are not capable of reliable producing greater than 16 bits of resolution at 100Hz bandwidth, the TG is a good match to the performance of most data acquisition systems.

Crossbow is planning to introduce an extremely high performance series sensor, the SM sensor. The SM sensor is designed specifically for ambient vibration monitoring, and features a 100 dB signal to noise ratio. The signal to noise ratio in excess of 100 dB allows small ambient vibrations to be measured even in areas of low seismic activity. The SM sensor, like the TG is fabricated with bulk micromachining, but it has a yet larger seismic mass than the TG and closed loop circuit electronics. One of the major challenges with the SM series is providing instrumentation electronics with signal to noise ratio commensurate with the sensor's capability.

CURRENT PROJECTS

UC Berkeley and Crossbow have actively participated in several research projects that have helped drive our understanding and development of these technologies. There have been two such demonstrations.

- Seismic Shaker Test of Apartment Building
- Tokachi Port Test of Liquefaction by Explosives

These tests have combined an LP class accelerometer and Renee MOPTEs as well as prototype CN4000 wireless capability.

The seismic shaker test was a series of shake tests conducted at the seismic shaker in Richmond, California. The test are parts of the CUREe-Caltech Tuck Under Apartment Building Experimentation. These types of buildings were severely and damaged during the 1994 Northridge Earthquake. The tests consists of dynamic tests on a full scale structure.

During the tests approximately 40 motes were used. Each mote had two axes of acceleration capability. Data was collected during the test and compared to wired accelerometers on the structure. The test demonstrated several key points. The wireless accelerometers were able to achieve a denser spatial instrumentation than wired accelerations. For example, in one test

30+ sensors were mounted to one sidewall of the structure. The wireless sensors were installed very quickly as compared to the centralized wired data acquisition system. The final point was that the quality of the data taken from the demonstration was equivalent to the data from the high performance wired sensors. (This is especially noteworthy sense the LP is our lowest cost sensor). In fact the wireless MOTE captured strong data that corresponded to the location of damage on one of sidewalls.

A second test was conducted at Tokachi Port, Hokkaido Japan. This test was blast-induced liquefaction. Buried explosives were placed in a large area of ground with trapped ground water. The explosives were set off to simulate liquefaction. The area was densely instrumented with a traditional wired sensor array. UC Berkeley and Crossbow also provided wireless instrumentation for this test. In this case approximately 30 motes with LP class accelerometers were provided. In this test as well, the benefits of wireless instrumentation was demonstrated. Quick set-up and a high-density of measurement was again shown. The benefits were even more evident because the area and hence the length of cable was longer in this test. Crossbow also provided about 18 traditional wired accelerometers and data logger systems. At this time, the data is being analyzed and it will be presented.

CONCLUSIONS

Structural health will provide compelling value for our preserving and extending the life of our infrastructure. Low-cost, smart sensors are key ingredient for wide spread adoption of structural health techniques. In this overview, we have summarized Crossbow's roadmap for structural and seismic sensing capabilities.

Wireless Communications between Standalone Semiactive Control Devices

L. Faravelli[1] and R. Rossi[2]

[1]Dept. of Structural Mechanics, Univ. of Pavia, Via Ferrata, 1 – 27100 Pavia, Italy
[2]Dept. of Electronics, Univ. of Pavia, Via Ferrata, 1 – 27100 Pavia, Italy

ABSTRACT

The design of a standalone semiactive control device has been one of the main research topics of the cooperation between the Departments of Structural Mechanics and Electronics at the University of Pavia.
The goal is pursued by adopting a programmable microprocessor which allows different classes of controllers (in particular fuzzy controllers) to be adopted, receiving feedback variables from sensors and determining the signals to be provided to the actuators.
The weak points of such an architecture are the wire connections between neighboring devices which provide a mutual feedback preventing contrasting control actions.
This contribution is an introduction to the design of a wireless connection between devices, which improve the reliability of the whole system.

1. INTRODUCTION

Fuzzy Logic theory (Passino and Yurkovich, 1998; Jang et al., 1997) has widely been proposed for the active control of structural systems (Casciati et al. 1996; Casciati and Yao, 1995; Faravelli and Yao, 1996). It easily allows the resolution of imprecise or uncertain information and, in particular, can handle structural nonlinearities.
The main advantages in adopting a fuzzy control scheme for Civil Engineering applications can be synthesized as follows:

- the implementation of fuzzy controllers is based on linguistic synthesis and, therefore, they are not affected by the choice of a specific mathematical model. As a consequence, the resulting fuzzy controller possesses inherent robustness;
- the uncertainties of input data from the external excitation and structural vibration sensors are treated in a much easier way by the fuzzy control theory than by the classical

control theory. Fuzzy logic, which is the basis of fuzzy controllers, intrinsically accounts for such uncertainties;

- the control action can be designed as a bounded function of the state variables. This provides an appropriate model for the actual behavior of the actuators.

2. SYSTEM DESIGN

When dealing with active controllers, the most general configuration for an engineering system makes use of a single controller block that processes a set of feedback variables and produces a control output, as in Figure 1. As opposed to classic control theory, fuzzy control theory does not provide the designer with a mathematical framework that allows him to simultaneously optimize the design of the controller and ensure its stability. Thus, partitioning the controller block into simpler parts can give the designer a deeper physical insight into the whole system operation.

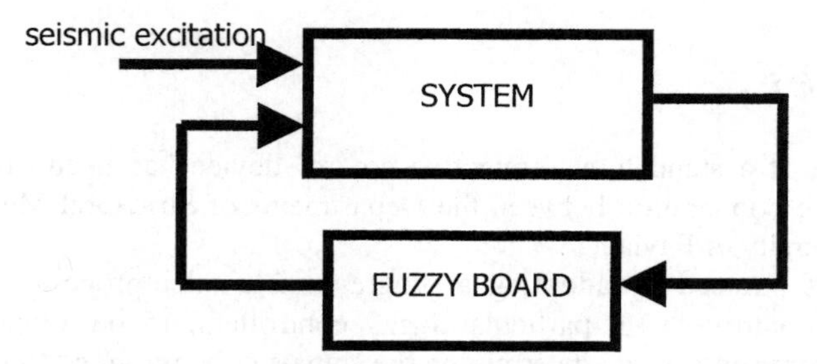

Figure 1. General control system configuration.

For this reason, when controlling a multi-degree-of-freedom structure, it may prove convenient to partition the system into smaller subsystems and, accordingly, split the controller into multiple controller blocks, each driving a single actuator, as shown in Figure 2. Also, cooperation among the blocks is a very important issue. Each controller block should be aware of how all the others are working, as failure to do so may result in worse performance or even instability.

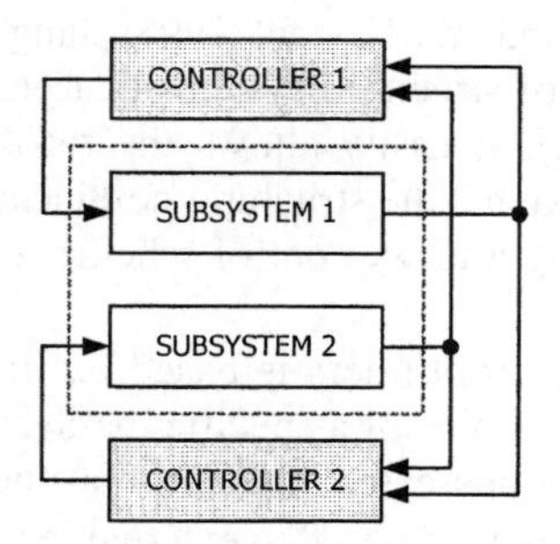

Figure 2. A control system after partitioning.

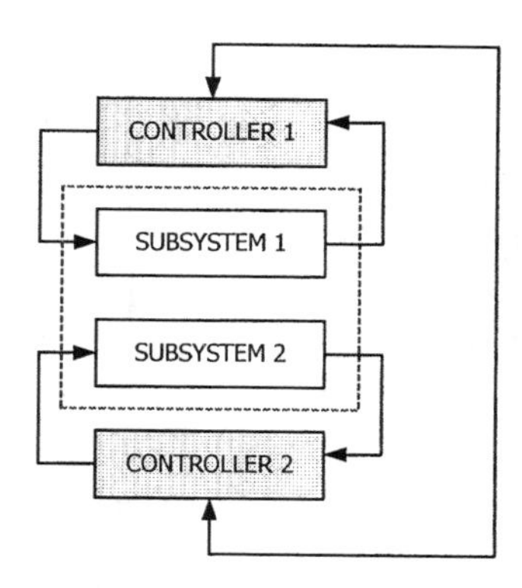

Figure 3. The same system of Figure 2 with a different wiring scheme. This requires fewer cables.

Communication between any two blocks can be implemented either in direct (i.e. by physically interconnecting them, like in Figure 3) or indirect form (i.e. by providing them with the same amount of information about the system motion, as shown in Figure 2).

This work has two main purposes: (1) to show how useful cooperation between controllers can be on a civil structure; (2) to suggest a possible way of implementing a controlling system based on cooperation and information exchange between blocks, in particular through a wireless interface.

First, we will deal with the best case, in order to show how effective cooperation may be. We will assume that we have two separate controllers sharing the same set of feedback variables, which means each block has a sort of indirect knowledge of what the other block is doing.

The example system we used is a three-story frame driven by two actuators located on the second and on the third floor, respectively. The system is controlled by two identical fuzzy chips, each connected to a single actuator. Both chips receive the velocities of the second and the third story, though differently weighted. Each of them calculates the control force to be introduced in the system by driving the actuator.

To speed up the design, it was deemed convenient to emulate the test frame by an electronic circuit. This allows fast testing of the controller, as there is no need to worry about possible damages to the test structure. This approach allows the performance of the system under design to be evaluated and optimized with no need for expensive and time-consuming laboratory tests on full-scale or scaled real structures. The equivalent circuit should be as simple as possible, to allow easy implementation and flexible use. An electronic circuit which emulates the original SDOF structural system was first conceived and implemented on a board (Casciati, Faravelli and Torelli, 1999). The same concept was then realized for the three-degree-of-freedom case.

Table 1. Fuzzy rules used in the simple example.

	NE	**ZE**	**PO**
NE	PL	PO	ZE
ZE	PO	ZE	NE
PO	ZE	NE	NL

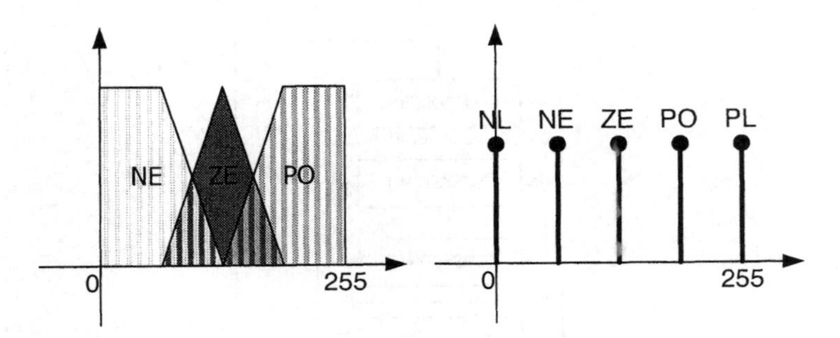

Figure 4. Fuzzy sets for the input and the output variables, respectively.

3. FUZZY CONTROLLER DESIGN

Each fuzzy controller has been implemented by using a commercially available fuzzy chip. The chip is arranged on a board to provide serial communication through the standard RS232-C interface, which allows easy downloading of the fuzzy controller code from a PC. The chip accepts up to eight inputs and computes up to four outputs by using up to 128 fuzzy rules.

In this work, each controller receives the same two velocities (v_2 and v_3) as inputs and produces the control signal to drive its own actuator. Figure 4 shows the three fuzzy sets adopted for the input variables and the five fuzzy sets used for the output variables. Table 1 reports the nine fuzzy rules used for the controller described in this paper. It is a very simple fuzzy project but this makes it a very robust one, as it does not target a specific mathematical model. So it can be used to control a relatively wide variety of systems, either linear, nonlinear or even hysteretic. The input velocities are differently weighted. In particular, the top actuator controller uses v_3 with weight 1 and v_2 with weight 0.6. The second controller uses 0.6 and 1 as weights for v_3 and v_2, respectively.

4. PERFORMANCE

The global controller was tested with a 1.2 Hz sinusoidal ground acceleration signal, which is close to the first resonance frequency of the structure. The amplitude of the input signal was set to produce a 5 cm oscillation on the third story in the uncontrolled situation. Figure 6 shows the performance of the controller when this kind of input signal is used to excite the structure. The plot is divided into three regions. The first one reports the behavior of the system in the uncontrolled case. In the middle region, the fuzzy controllers are not communicating (Figure 3). The amplitude of the third story oscillations drops from 5 cm to 7 mm. The improvement over the uncontrolled situation is about 17 dB.

Finally, the third region shows the behavior of the system when the controllers are communicating. A further reduction of the structural oscillations to about 4.4 mm is experienced, resulting in a total 21 dB improvement over the uncontrolled case.

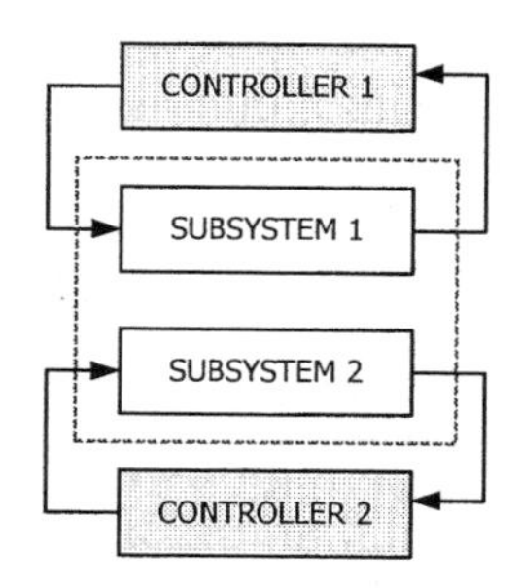

Figure 5. Two concurrent controllers that do not communicate with each other.

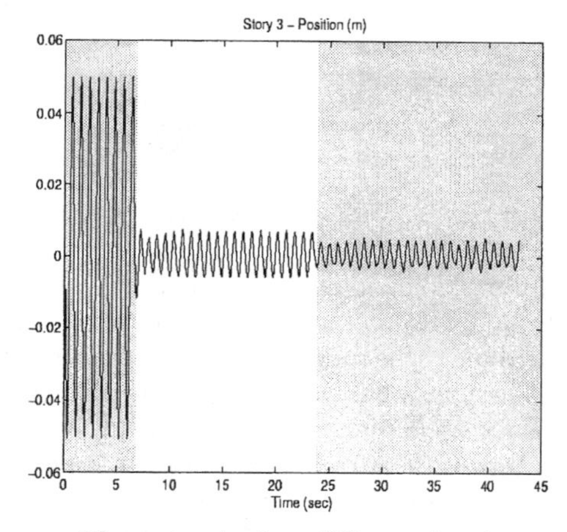

Figure 6. Performance of the system in three different situations, as explained in the text.

5. WIRELESS IMPLEMENTATION ISSUES

Once the advantages of cooperation are clear, we will now focus on a possible way to implement such a controller.

Civil structures usually present instrumentation and wiring problems, especially due to long distance and environmental exposure. To solve this, a wireless network could be the ideal solution (Figure 7). Wireless technology is developing at a fast pace nowadays and will keep growing in the years to come.

Several off-the-shelf wireless communication integrated circuits are available today, each one posing different issues about band, power, synchronization and overall system complexity.

We already have a standalone fuzzy controller board capable of controlling civil structures and it is very easy to connect two or more of them with a simple cable. The problem we plan to solve in the near future is how to remove the cable link and replace it with a wireless connection.

In our case, we are talking about connecting two controllers. This means a radio modem could be the best solution. Several radio modem integrated circuits are available today, with plenty of variations. In the simplest case, they expose an interface that is identical to the standard RS-232 serial interface. Our board already implements such an interface, thus making it easy to implement such a communication scheme on our fuzzy controller board.

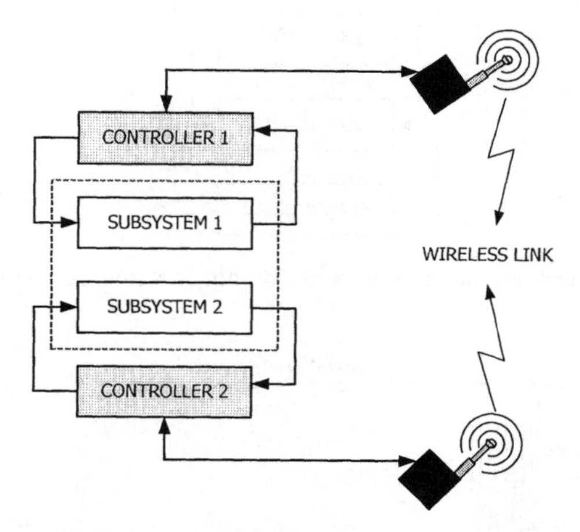

Figure 7. The system of Figure 3 with a wireless link between the two controllers.

No particular synchronization problems may occur when only two controllers are involved. In case more than two controllers must communicate with each other, each controller might be assigned an address and then each controller would start transmitting as soon as the previous one finished. Let us assume we have a device that enables wireless communication at 19200 baud rate. Let us also assume each block transmits two bytes (i.e., the address of the sender and the information). If the rate at which this information has to be transmitted is 100 Hz and we have N devices, then one sampling interval must allow for 2N bytes. If each byte requires a start bit and a parity bit, then $20N/19200 < 1/100$ gives the upper bound for the number of devices connected together. In this case we have $N <= 9$. This simple round robin synchronization mechanism allows 9 devices to communicate with each other at 100 Hz if the baud rate is 19200. There is even some margin for some guard bits between data packets. Now, if we decide that 8 devices are enough, of the two bytes we transmit only three bits are necessary for the sender's address. The remaining 13 bits can hold the data.
Also, it could be decided that the exchanged information need not be transmitted at a rate of 100 Hz. A lower sampling rate might be enough for this type of data. In this case, even more devices could be connected together. In any case, the actual configuration will depend on the particular application.
In order to prevent the system from halting in case of poor transmission quality, as well as to start the system, a little amount of watchdog timer logic is necessary yet very straightforward. At its simplest, the block with address 0 could be instructed to start transmitting after a given timeout. However, with the microcontroller we use, it is very easy to implement a more sophisticated watchdog algorithm, possibly capable of detecting if any of the N devices is up and working or if it needs to be bypassed.
A more sophisticated way to handle communications could be through an RF transceiver, which can still be viewed as a radio modem though not supporting the RS-232 serial interface. For example, the TRF6900 allows transmission in the ISM (Industrial, Scientific, Medical) unlicensed band (i.e., 868 – 870 MHz in Europe, 902 – 928 MHz in the USA). This integrated circuit allows a 20 MHz serial clock frequency (i.e., equivalent baud rate) to be

used. With this chip, much higher performance can be achieved compared to the simpler solution outlined above. As to synchronization, a similar scheme may be adopted. The resulting maximum number of devices is greatly raised, as well as the amount of data per sampling interval and/or the transmission rate.

6. CONCLUSIONS

In this paper, the benefits of concurrent control have been first stressed out, with regard to the development of a wireless communication interface between two or more control devices. Next, a few considerations about the design of the wireless interface have been made with a view to future implementation.

ACKNOWLEDGEMENT

This research was supported by grant I/R/187/01.

REFERENCES

Casciati F., Faravelli L. and Torelli G., 1999, A Fuzzy Chip Controller for NonLinear Vibrations (accepted for publication in Nonlinear Dynamics).

Casciati F., Faravelli L. and Yao T., 1996, Control of Nonlinear Structures Using the Fuzzy Control Approach. Nonlinear Dynamics, 11, 171-187.

Casciati F. and Giorgi F., 1996, Fuzzy Controller Implementation. Proceedings of 2nd International Workshop on Structural Control. IASC, Hong Kong, 119-125.

Casciati F. and Yao T., 1995, Comparison of Strategies for the Active Control of Civil Structures, Proceedings of 1st world Conference on Structural Control, IASC, Los Angeles, Vol.I, WA1-3.

Faravelli L. and Yao T., 1996, Use of Adaptive Network in Fuzzy Control of Civil Structures, Microcomputers in Civil Engineering, 11 (1), 67-76.

Jang J.S.R., Sun C.T. and Mizutani E., 1997, Neuro-Fuzzy and Soft Computing, Prentice Hall Inc.

Passino K.M. and Yurkovich S., 1998, Fuzzy Control, Addison Wesley Longman Inc.

STMicroelectronics, 1996, Fuzzystudio™ 2.0 User Manual.

Sedra A.S. and Smith K.C. , 1991, Microelectronic Circuits, Third Edition, Saunders College Publishing, Philadelphia.

R-SHAPE: A Real-Time Structural Health and Performance Evaluation System

Wilfred D. Iwan

Earthquake Engineering Research Laboratory, California Institute of Technology, Pasadena, CA, USA.

ABSTRACT

This paper describes a recent development in structural health and performance monitoring referred to as the Caltech Real-Time Structural Health and Performance Evaluation (R-SHAPE) System. This system is installed in the Millikan Library Building on the campus of the California Institute of Technology in Pasadena, California. It provides true-real time data over the Internet to any location in the world and is integrated with various real-time health and performance monitoring tools.

1. INTRODUCTION

A variety of different systems have been proposed and implemented for studying the health and performance of building structures. These involve the use of sensors ranging from accelerometers to strain gages to displacement meters. Some of these systems have the potential for near real-time data acquisition, but most involve post processing of the data after an event is recorded in order to determine structural performance.

Real-time structural health monitoring has many advantages over post-processed monitoring. First, real-time monitoring provides a basis for rapid decision making under adverse conditions. There is increasing demand for very quick response to natural disasters such as extreme winds and earthquakes as well as man made disasters such as terrorism attacks and technological accidents. In order to be most effective, this response often needs to be initiated at the height of the crisis. Real-time health and performance data can help to insure that decisions are made with appropriate information. Indeed, many decisions might be automated if real-time data is available.

A second more subtle motivation for real-time monitoring of structural health and performance is that it can result in increased public awareness, understanding, and acceptance of monitoring technology. Public support and even public demand can be extremely important factors in driving the development of refined systems and improved technologies. This has been proven by past experience with ground motion monitoring of earthquakes.

2. AN OVERVIEW OF REAL-TIME MONITORING IN THE US

Real-time earthquake monitoring and warning systems for urban regions were first proposed for seismically active regions of the US in 1979 (Iwan, 1979). However, it was not until the early 1990's that such systems began to be deployed. The Caltech-USGS Broadcast of Earthquakes (CUBE) system was built upon existing sensor technology but employed advances in data transmission and processing to create a radical new earthquake monitoring system. This system was capable of determining an earthquake epicenter location and magnitude within several tens of minutes after an event. Information was broadcast by pager to subscribers nationwide and was also made available in graphical form to subscribers using personal computers. The CUBE system has been, and continues to be used by a broad spectrum of users for timely decision making following an earthquake. Decisions range from whether to stop or re-route railroad rolling stock to how to deploy utility maintenance crews. Only ground motion sensors were employed in the CUBE system.

Following the Northridge Earthquake of 1994, a new ground motion monitoring system grew out of the CUBE system. The TriNet system is a joint effort between Caltech, the US Geological Survey (USGS), and the California Geological Survey (CGS). Its development was supported by the Federal Emergency Management Agency (FEMA) under Stafford Act funding for Northridge earthquake mitigation activities. The goal of the TriNet system is to provide rapid (several minutes) broadcast of earthquake location and shaking intensity maps. The shake maps represent a significant advancement over the simple magnitude measure provided by CUBE. The success of the TriNet program in Southern California has prompted its extension to Northern California and the program has been renamed the California Integrated Seismic Network (CISN). This program is presently focused primarily on ground motion monitoring.

In 1999, the USGS undertook development of the Advanced National Seismic System (ANSS). This is a very ambitious undertaking which has not yet been fully funded. It will ultimately involve the installation of approximately 6000 new instruments, about one-half in buildings. Most of the ground motion instruments will be tied into near real-time systems such as TriNet, and some of the building monitoring systems will likely be near real-time.

Also in 1999, the Consortium of Organizations for Strong-Motion Observations Systems (COSMOS) was formed as a non-profit US public benefit corporation. This organization was formed by a core group of state and federal agencies involved in strong-motion monitoring. The goal of the organization is to promote the advancement of strong-motion measurement on the ground and in structures, and encourage the rapid, convenient, and responsive distribution of strong-motion data according to standards developed by the Consortium. The organization has developed standards and guidelines for deployment of instruments for both ground motion and building monitoring. However, it is not yet taken steps to promote real-time structural monitoring.

Last year, a Tri-Net station was installed in the Millikan Library building on the campus of the California Institute of Technology (Heaton, 2001). This site has the station designation MIK. Data from this site is treated in exactly the same manner as that from any other TriNet station. It is available with ground motion data in the archive of the Southern California Network. 80 sps data is archived on a continuous loop for one week, and 20 sps data is archived indefinitely. Additional TriNet stations are being installed in other buildings on the Caltech campus. Some potential advantages of the TriNet station approach are: 1) the data is readily assessable, 2) the data is continuously maintained, 3) there is no distinction between building and ground motion data in terms of resolution, formatting, or archiving, and 4) both ground motion and building data may be studied in a unified manner. The high gain and low noise threshold of the 24-bit TriNet station enables the examination of building behavior not previously studied. For example, a small but systematic variation is observed in mean of the ambient acceleration response over a one-day period, and the there is a small but quantifiable change in the natural periods of the building during rainy periods. Such observations may ultimately lead to a greater understanding of the parameters that affect structural response behavior.

Based on the extension of the TriNet system to a building structure, a new near real-time monitoring system was installed in Millikan Library in 2001. This system is referred to as the Caltech Online Monitoring and Evaluation Tools (COMET) system (Beck, 2001). This system is capable of on-demand downloading of data from the MIK station. Data at 80 sps is maintained on the system for three months. The system is also planned to have a number of data analysis tools. The COMET system may be accessed at http://www.comet.caltech.edu.

2. THE R-SHAPE SYSTEM

Early in 2002, a true real-time monitoring system was installed in the Caltech Millikan Library building. The system is called R-SHAPE for Real-Time Structural Health and Performance Evaluation. This system is the result of a joint effort by Caltech, the USGS, and Digitexx Data Systems, Inc. It is based on current sensor technology, but provides an excellent test bed for new sensor and analysis software development, as well as for the design and application of decision-making tools.

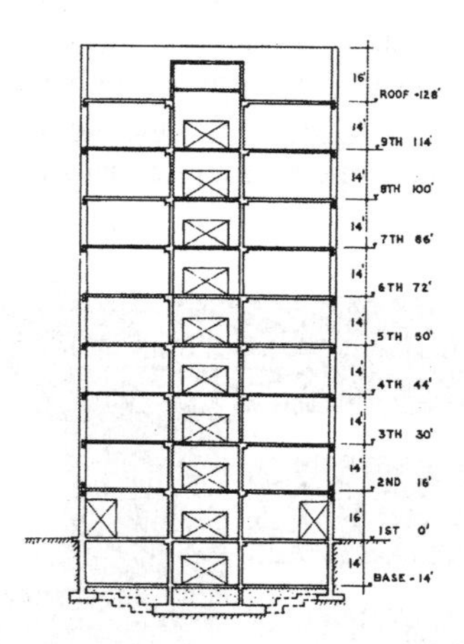

Figure 1. N-S Elevation view of Millikan Library Building.

Millikan Library is a nine-story reinforced concrete frame building that was dedicated in 1966. Figure 1 shows an elevation view of the building. The building has been well instrumented since its construction and has recorded motions during all major Southern California earthquakes. Millikan Library has been extensively modeled and studied by researchers worldwide. The building is also routinely shaken by a rotating eccentric mass exciter located on its roof. Its dynamic properties are very well known and well understood.

The monitoring system is built upon a commercially available Kinemetrics Whitney Box type of installation. This installation is typical of that used by the USGS and CGS throughout California and elsewhere in United States. There are 36 channels of Forced Balance Accelerometer (FBA) data fed to a triggered recorder that locally records all events exceeding a specified level. The recorded data may be downloaded by phone line following an earthquake event. Figure 2 indicates the location of the accelerometers within the building.

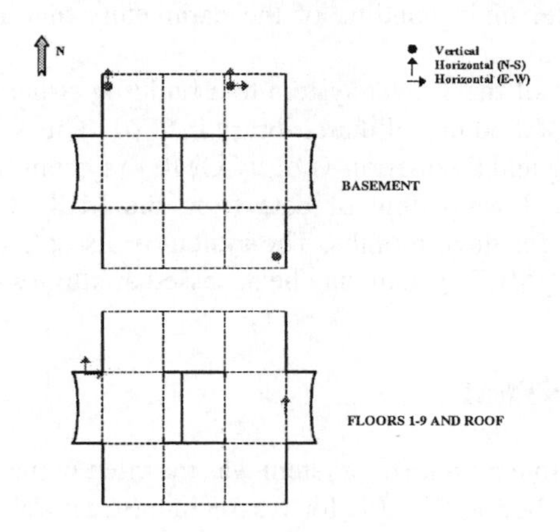

Figure 2. Layout of location of accelerometers in Millikan Library

The new real-time monitoring system uses a data feed from the existing FBA sensors with 16-bit A to D conversion at 100 sps. The R-SHAPE system does not interfere with normal functioning of the Whitney Box system. Streaming digital data is transferred to a LAN using TCP-IP protocol. This data is sent to a remote server where it is made available over the Internet. The web address is: www.R-SHAPE.caltech.edu. The web page shows real-time streaming data for acceleration, real-time Fourier Transform of acceleration, and a live image of the system itself (see Figure 3).

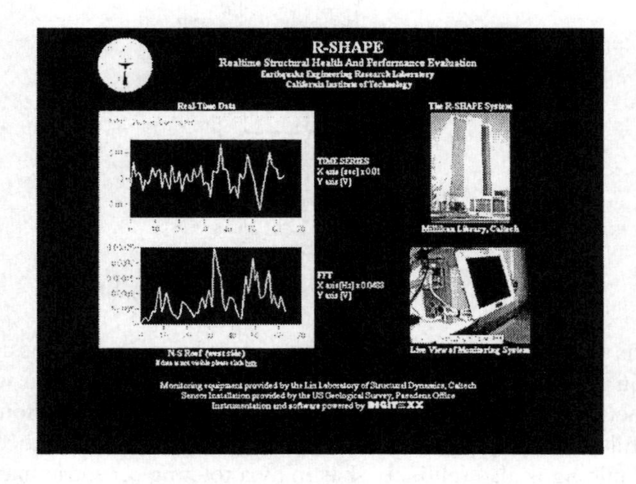

Figure 3. R-SHAPE web site opening screen

Client software is available for use on a user's computer that provides real-time graphical views of streaming data for all 36 channels (see Figure 4). With this software, the data may be stored locally for later analysis. Software is currently available for several forms of real-time analysis including continuous displacement, Fourier Transform, and continuous Response Spectrum (see Figure 5). These are updated on a real-time basis. Software can also be used to automatically sound an alert if certain pre-defined response limits are exceeded.

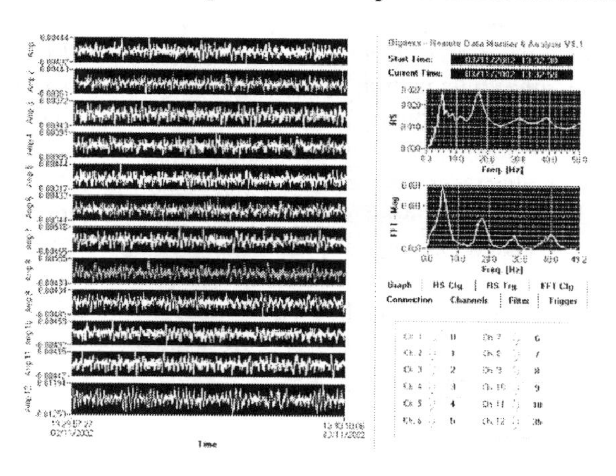

Figure 4. Client software available for on-line analysis of R-SHAPE data. 12 of 36 channels of streaming acceleration data plus Response Spectrum and FFT

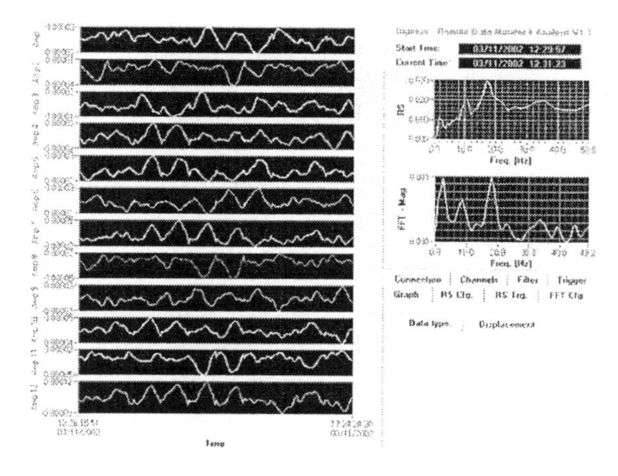

Figure 5. Client software available for on-line analysis of R-SHAPE date. 12 of 36 channels of streaming displacement data plus Response Spectrum and FFT.

3. APPLICATIONS OF THE R-SHAPE CONCEPT

The R-SHAPE system can readily be used for real-time monitoring of a number of important performance indicators. These include: 1) natural frequencies of vibration, 2) mode shapes of vibration, 3) peak inter-story drift, 4) peak inter-story shear, and 5) inter-story hysteresis diagrams. The accuracy with which these indicators can be determined is dependent upon the size of the data sample employed and the algorithms used. However, a distinct advantage of continuous real-time monitoring is that accurate baseline information and models can be evaluated and updated continuously. This makes it much easier to rapidly identify changes in performance indicators.

Based on the real-time performance indicators that may be presented, a wide range of decisions can be made in near real-time. Long-term changes in building properties may indicate general building deterioration or even "healing". For example, it has been previously observed that the small amplitude natural frequency of the lowest response mode of Millikan Library decreases immediately following a significant earthquake, but recovers again over an extended period of time following the event.

If measured loads or displacements exceed their design limits, this is an indication of possible structural damage. If significant short-term changes are observed in structural properties, it may be concluded that there is probable structural damage. By observing various inter-story indicators, it may be possible to identify the approximate location and degree of damage. This information could provide the basis for important decisions such as whether the building should be evacuated for occupant safety following an earthquake.

4. CONCLUSIONS

Based on the experience of the author, the following conclusions have been reached:

1. Real-time structural health and performance monitoring of buildings is a reality and the concept is rapidly gaining support.
2. The potential benefits of real-time building monitoring for decision making by building owners, emergency managers and responders, and engineers are very significant.
3. These benefits will only be fully realized as the technology is more widely implemented, exercised, and improved.

ACKNOWLEDGEMENT

The author gratefully acknowledges the support and assistance of Prof. James Beck (COMET), Prof. Thomas Heaton (MIK), Mr. Costin Radulescu (Digitexx), and Dr. Erdal Safak (USGS).

REFERENCES

Beck, J.L. (2001) Personal Communication.

Heaton, T.H. (2001) Personal Communication.

Iwan, W.D. (1979) California's San Andreas Fault - New System Might Give Residents and Industry Up to a Minute's Notice of a Major Quake. *The Christian Science Monitor*, June 6, 1979.

Iwan, W.D. (1979) Earthquake Warning System Could Help Lessen Damage. *The Los Angeles Times*, June 29, 1979.

USGS (1999) An Assessment of Seismic Monitoring in the United States; Requirement for an Advanced National Seismic System, Circular 1188

Damage Identification for Structures Using Hilbert-Huang Spectral Analysis

Jann N. Yang, Silian Lin and Shuwen Pan

Department of Civil and Environmental Engineering, University of California, Irvine, CA92697, USA.

ABSTRACT

In this paper, applications of the Hilbert-Huang spectral analysis for the quantitative system identification and damage detection and evaluation are described and presented. In particular, a methodology is demonstrated using the ASCE structural health monitoring benchmark structure. The structural parameters, including stiffness and damping, before and after damage are identified based on the noisy measurements of acceleration responses. Then the location and severity of the damage are assessed by a comparison. Simulation results demonstrate that the accuracy of the methodology presented in identifying the structural damages is very plausible, and it represents a viable damage identification approach for linear structures.

1. INTRODUCTION

For structural health monitoring, appropriate data analysis techniques are needed to interpret the measured data and to identify the state of the structure and its damage after a severe event, such as earthquake. Various data analysis techniques have been available in the literature for the health monitoring of linear structures [e.g., Chang (1997, 1999, 2001)]. In order to facilitate a comparison of available techniques on the common basis, the ASCE Task Group on Structural Health Monitoring has established a benchmark problem [Johnson et al (2001)].

Recently, a method of decomposing a signal in the frequency-time domain using the Hilbert transform has been proposed by Huang et al [1998,1999]. This method is to decompose the signal into intrinsic mode functions (IMFs) using the Empirical Mode Decomposition (EMD) method, where each intrinsic mode function (IMF) admits a well-behaved Hilbert transform. Then, the Hilbert transform is applied to each intrinsic mode function to obtain a decomposition of the signal in the frequency-time domain. Such an approach is referred to as the Hilbert-Huang spectral analysis (HHSA) and it is applicable to nonstationary and nonlinear structural responses. Based on the Hilbert-Huang spectral analysis (HHSA), methodologies have been developed for the quantitative system identification and damage detection of linear structures [e.g., Yang et al (1999, 2000a,b,c,d, 2001, 2002a,b)]. In this paper, the HHSA and its applications to the system identification and damage detection of linear structures will be described first. Then it will be applied to the ASCE benchmark structure to demonstrate its ability to detect the structural damage.

2. HIBERT-HUANG SPECTRAL ANALYSIS (HHSA) AND ITS APPLICATIONS TO SYSTEM IDENTIFICATION

Suppose x(t) is a general measured signal. The Hilbert transform (HT) of x(t), denoted by $\tilde{x}(t)$, is given by

$$\tilde{x}(t) = HT[x(t)] = \int_{-\infty}^{+\infty} \frac{x(\tau)}{\pi(t-\tau)} d\tau \quad (1)$$

where $\tilde{x}(t)$ can be computed numerically using the efficient Fast Fourier transform [Huang et al (1998, 1999)]. The analytical signal Y(t) of x(t) is expressed as

$$Y(t) = x(t) + \underline{i}\,\tilde{x}(t) = A(t)\, e^{\underline{i}\,\theta(t)} \quad (2)$$

in which A(t) is the instantaneous amplitude, $\theta(t)$ is the instantaneous phase angle, and $\underline{i} = (-1)^{1/2}$. The instantaneous frequency $\omega(t)$ of the signal is obtained as

$$\omega(t) = d\theta(t)/dt \quad (3)$$

From Eqs.(1)-(3), there is only a single frequency ω at any time t, if the general signal x(t) is processed through the Hilbert transform. On the contrary, for a general signal x(t) at any time t, there is a distribution of frequencies at that time rather than just a single frequency. Consequently, to obtain a meaningful decomposition of a signal in the frequency-time domain, it was proposed by Huang et al (1998) to first decompose the signal x(t) into m Intrinsic Mode Functions (IMFs) that admit well-behaved Hilbert transform using the Empirical Mode Decomposition (EMD) method.

The procedure of the EMD method is to construct the upper and lower envelopes of the signal by spline-fitting and the average (mean) of both envelopes are computed. Then, the signal is subtracted from the mean, referred to as the sifting process. By repeating the sifting process until the resulting signal becomes monocomponent, i.e., one up-crossing (or down-crossing) of zero will result in one local peak (or trough). Such a monocomponent signal admits a well-behaved Hilbert transform and it is referred to as an IMF. The original signal is then subtracted from the IMF and the sifting process is applied to the remaining signal to obtain another IMF. The sifting process is repeated to obtain m IMFs, i.e.,

$$x(t) = \sum_{j=1}^{m} c_j(t) + r_m(t) \quad (4)$$

in which $c_j(t)$ ($j = 1, 2, \ldots, m$) are IMFs of the measured signal x(t) and $r_m(t)$ is the residue that can be the mean trend or a constant. The entire process above is referred to as the EMD method. It has been shown by Huang et al (1998,1999) that the characteristics of the signal can be extracted through the behavior of the Intrinsic Mode Functions (IMFs), and the EMD is applicable to nonstationary or nonlinear signals. Based on the EMD approach described above, the first IMF has the highest frequency content of the signal. During the EMD process, a specified frequency, referred to as the intermittency frequency ω_{int}, can be imposed so that the resulting IMF will have frequencies higher than ω_{int}. This is accomplished by removing data, which have frequencies lower than ω_{int}, from the IMFs. Finally, Eq.(4) is substituted into Eq.(1), and the Hilbert transform is applied to each IMF to form a meaningful frequency-time decomposition $A(t, \omega; x)$ of a general signal $x(t)$, referred to as the Hilbert-Huang spectral analysis.

Suppose x(t) above is the response of a linear structure, either the free vibration response or the response due to white noise excitations. The response due to the jth mode, referred to as the jth modal response and denoted by $x_j(t)$, can be obtained using the EMD method with appropriate intermittency frequency or using the EMD method and band-pass filter [e.g., Yang et al (2001)]. Hence, Eq.(4) can be expressed as

$$x(t) = \sum_{j=1}^{n} x_j(t) + \sum_{j=1}^{m-n} c_j(t) + r_m(t) \tag{5}$$

Then, the Hilbert transform can be applied to each modal response $x_j(t)$ ($j = 1, 2, \ldots, n$) to obtain the modal frequencies, damping ratios, mode shapes, and stiffness and damping matrices of the structure.

Based on the Hilbert-Huang analysis (HHSA) above, methods have been presented to identify all the natural frequencies and damping ratios of linear structures with only one measurement (sensor) of the noise-polluted acceleration response [Yang et al (1999, 2000a, b)]. When the acceleration responses at all DOFs are measured, the eigenvalues and eigenvectors (either real or complex), and the stiffness and damping matrices of linear structures can be accurately identified [Yang et al (2000a, b)]. Accurate estimation of dampings for tall buildings can be quite challenging, in particular the damping ratios of higher modes. Nature frequencies and damping ratios of in-situ tall buildings can be identified accurately using the HHSA above and a single measurement of the acceleration response under ambient wind vibration. This has been demonstrated in [Yang et al (2000c, d)] using a wind-excited benchmark tall building.

When the measured data include the damage events, the EMD method can be used to identify the time at which damage occurs and the damage location in the structure [Yang et al (2001)]. For the ASCE benchmark building, when all braces in the first story unit are removed

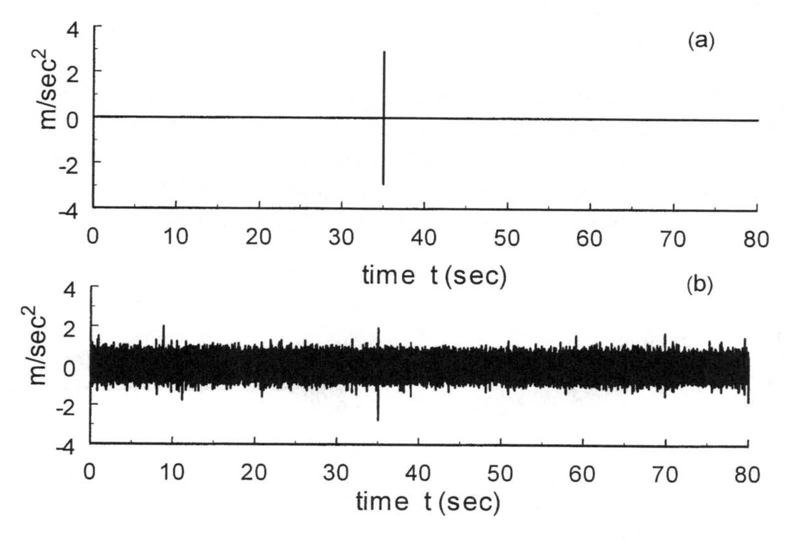

Fig. 1: Damage spikes in first floor acceleration; (a) without noise, (b) with noise.

at 35 second, the EMD method is capable of detecting the damage spike from the acceleration response data of the first floor as shown in Fig.1(a) and 1(b), respectively, for data without and with noises. When the noise level is high, however, the damage spike may be merged in the noise and hence it cannot be detected. In this case, the HHSA can be applied to the first modal response in Eq.(5) and the maximum amplitude in the time-frequency domain can be plotted as shown in Fig.2. As observed from Fig.2, damage clearly occurs at 35 second.

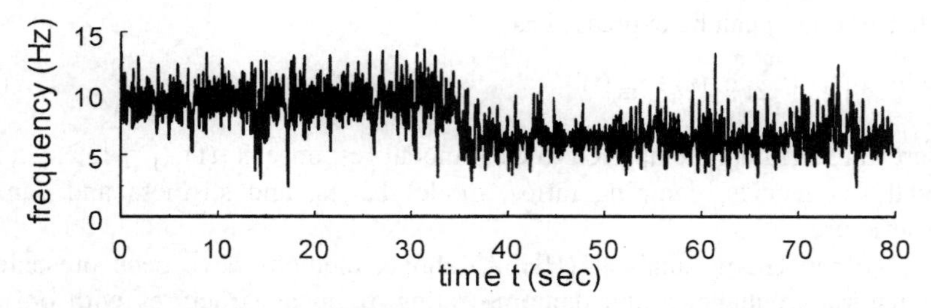

Fig. 2: Maximum amplitude of frequency-time decomposition of first modal response

3. DAMAGE EVALUATION FOR ASCE BENCHMARK STRUCTURE

The ASCE benchmark structure is a 4-story building model and the corresponding analytical models have been developed. The excitation forces are unknown white noise processes applied to each floor of the building. Damage is introduced into the structure by removing braces in the analytical model. Only acceleration measurements are available, which are also polluted by uncorrelated noises. Details of the benchmark problem are given in [Johnson, et al (2001)]. The cross-correlation function between the rth floor acceleration $\ddot{x}_r(t)$ and that of the sth floor $\ddot{x}_s(t)$, denoted by $R_{rs}(\tau)$, can be obtained as

$$R_{rs}(\ ) = E[\ddot{x}_r(t)\ddot{x}_s(t+\ )] = \sum_{j=1}^{n} R_{rs,j}(\ );\ R_{rs,j}(\tau) = \varphi_{rj}\alpha_{sj}\, e^{-\zeta_j\omega_j}\ \sin(\omega_{dj}\tau + \ _{sj}) \quad (6)$$

in which $R_{rs,j}(\tau)$ is the contribution from jth mode, ϕ_{rj} is the rth element of the jth modal vector, ω_{dj} is the jth damped modal frequency, θ_{sj} is the phase lag, and α_{sj} is a positive constant. A representative cross spectral density function and the corresponding cross correlation function of the 4th and 1st floors of the undamaged structure are shown in Fig.3.

The EMD method and band-pass filters have been used to extract each modal contribution

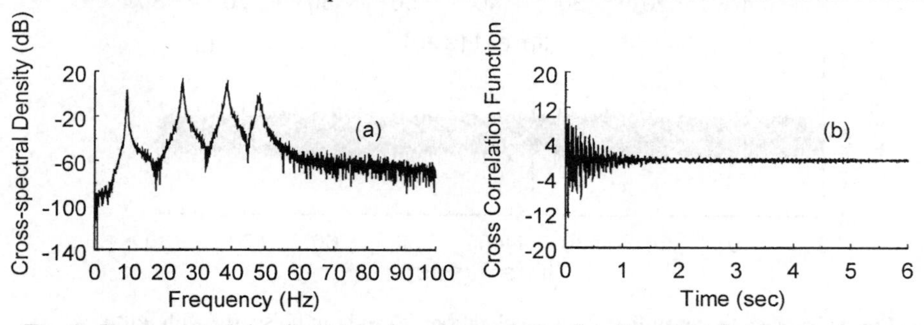

Fig. 3: Cross-spectral density, (a), and cross-correlation function, (b), of undamaged structure.

in Eq.(6) and the Hilbert transform has been applied to obtain the instantaneous amplitudes and phase angles [Yang et al (2002a,b)]. The identified natural frequencies and damping ratios for the undamaged and damaged structures (damage patterns 1 and 2 of Case 1) are presented in Table 1 as denoted by "ID". Also shown in Table 1 are the theoretical values, denoted by "TH", for comparison. The theoretical damping ratio for each mode is 1%. For the damage pattern 1, all braces in the first story unit are removed, whereas for the damage pattern 2, all braces in both the first and third story units are removed. As observed from Table 1, the identified natural frequencies and damping ratios are quite accurate. Assuming that the mass matrix is known, the stiffness matrix **K** and damping matrix **C** have been estimated. The stiffness in each story unit is computed based on the shear-beam model. The identified results for the stiffness in each story unit, denoted by k_1, k_2, k_3 and k_4, respectively, are presented in Table 2. Also shown in Table 2 are the theoretical values for comparison. From the identified stiffness in each story unit presented in Table 2, the damage locations and damage severities have been detected and identified quite accurately.

Table 1: Identified Natural Frequencies And Damping Ratios

Mode	Undamaged			Damage Pattern 1			Damage Pattern 2		
	Frequency (Hz)		Damping (%)	Frequency (Hz)		Damping (%)	Frequency (Hz)		Damping (%)
	TH	ID	ID	TH	ID	ID	TH	ID	ID
1	9.41	9.39	1.09	6.24	6.25	1.17	5.82	5.83	1.01
2	25.54	25.41	1.16	21.53	21.51	1.04	14.89	14.88	1.03
3	38.66	38.59	1.05	37.37	37.35	0.88	36.06	36.02	1.10
4	48.01	48.13	1.03	47.83	47.81	1.12	41.35	41.37	1.00

Table 2: Stiffness Parameter of Each Story Unit For Case 1

Story Stiffness (MN/m)	Theoretical Values			Identified Values		
	No Damage	Damage Pattern 1	Damage Pattern 2	No Damage	Damage Pattern 1	Damage Pattern 2
k_1	67.90	19.68	19.68	67.71	19.02	20.26
k_2	67.90	67.90	67.90	68.58	68.38	67.36
k_3	67.90	67.90	19.68	68.25	68.38	19.21
k_4	67.90	67.90	67.90	67.35	66.91	68.42

4. CONCLUSIONS

Applications of the Hilbert-Huang spectral analysis for the quantitative system identification and damage detection and evaluation have been described and presented. A methodology has been applied to the ASCE benchmark building model. All the natural frequencies and damping ratios of the structure can be identified with only one measurement of the noise-polluted acceleration. When the acceleration responses at all DOFs are measured, the modal matrix, stiffness matrix and damping matrix can be identified. Simulation results demonstrate that the methodology presented is quite accurate in identifying the structural damages, including the damage locations and severities. It is quite effective for the system identification and damage detection of linear structure.

ACKNOWLEDGEMENTS

This research is supported by the National Science Foundation through Grant No. CMS-98-07855.

REFERENCES

Chang, F.-K., ed. (1997, 1999, 2001). *Structural Health Monitoring*, Proc. 1st, 2nd and 3rd International Workshop on Structural Health Monitoring, Stanford University, Stanford, CA. CRC Press.

Huang, N.E., Shen, Z., Long, S.R., Wu, M.C. and Shih, H. H., et al. (1998). "The Empirical Mode Decomposition and Hilbert Spectrum for Nonlinear and Nonstationary Time Series Analysis". *Proc. Royal Society of London Series*, A454, 903-995.

Huang, N.E., Shen, Z. and Long, S.R. (1999). "A New View of Nonlinear Water Waves: The Hilbert Spectrum". *Annu. Rev. Fluid Mech.*, 31, 417-457.

Johnson, E.A., Lam, H.F., Katafygiotis, L.S., and Beck, J.L. (2001). "A Simulated Data Benchmark Problem in Structural Health Monitoring". *Proc. 3rd Inter. Workshop on Structural Health Monitoring,* Stanford, CA. 467-477; also http://wusceel.civil.wustl.edu/ asce.shm/.

Yang, J. N, and Lei, Y. (1999), "Identification of Natural Frequencies and Damping Ratios of Linear Structures Via Hilbert Transform and Empirical Mode Decomposition", *Proc. of IASTED International Conf. on Intelligent Systems and Control*, 310 – 315, Santa Barbara, CA.

Yang, J.N., and Lei, Y. (2000a). "System Identification of Linear Structures Using Hilbert Transform and Empirical Mode Decomposition", *Proc. of 18th International Modal Analysis Conference: A Conference on Structural Dynamics,* Vol.1, pp.213-219, Feb. 7-10, San Antonio, TX, Paper No. 256, Society for Experimental Mechanics, Inc., Bethel, CT.

Yang, J.N., and Lei, Y. (2000b). "Identification of Civil Structures with Nonproportional Damping" *Smart Structures and Materials 2000: Smart Systems for Bridges, Structures & Highways,* Proc. of SPIE, Vol. 3988, 284-294. Newport Beach , CA.

Yang, J.N., and Lei, Y. (2000c). "Identification of Tall Building Using Noisy Wind Vibration Data". *Advances in Structural Dynamics,* Vol.II, 1093-1100. Elsevier Science Ltd., *Proc. International Conf. On Advances in Structural Dynamics,* Hong Kong.

Yang, J.N., and Lei, Y. (2000d), "Parametric Identification of Tall Buildings Using Ambient Wind Vibration Data", *Proc. 8th ASCE Specialty Conf. on Probabilistic Mech. and Struct. Reliability*, Paper No. PMC2000-076, CD RAM 6 pages, Univ. of Notre Dame, IND.

Yang, J.N., Lei, Y. and Huang, N.E. (2001). "Damage Identification of Civil Engineering Structures Using Hilbert-Huang Transform". *Proc. 3rd Inter. Workshop on Structural Health Monitoring,* 544-553. Sept. 11-13, Stanford, CA.

Yang, J.N., Lin, S. and Pan, S. (2002a). "Damage Identification of Structures using Hilbert-Huang Spectral Analysis", *Proc. 15th ASCE Engrg. Mechanics Conf.* (CD-ROM), New York.

Yang, J.N., Lin, S. and Pan, S. (2002b). "Damage Detection of A Health Benchmark Building Using Hilbert-Huang Spectral Analysis", *Proc. of Advances In Building Technology* (*ABT2002*), December 4-6, Hong Kong.

Fiber Optic Sensors in Civil Structures

Farhad Ansari[1]

Department of Civil & Materials Engineering, University of Illinois at Chicago, Chicago, Illinois, USA

ABSTRACT

A brief introduction into the theories, principles, and applications of fiber optic sensors is provided in this article. Optical fibers are geometrically adaptable and they can be used either as embedded or adhered sensors. Some applications to structural engineering are presented, especially in applications pertaining to cracking and deformations of materials.

1. INTRODUCTION

Fiber optic sensors are usually classified in terms of the transduction mechanism, and they are either intensity, wavelength, or phase based in construction. The measurand, i.e. strain, induces a change in the intensity, wavelength, or phase of the light propagating in the sensing region of the optical fiber. This change, in turn, is related to the measurand through the interrelationship between the optical phenomena and the specific measurand. The choice of approach usually depends on the specific application and requirements of the sensor. For instance, sensors can be categorized based on the application, or the transduction mechanism. These classifications are diagrammatically depicted below:

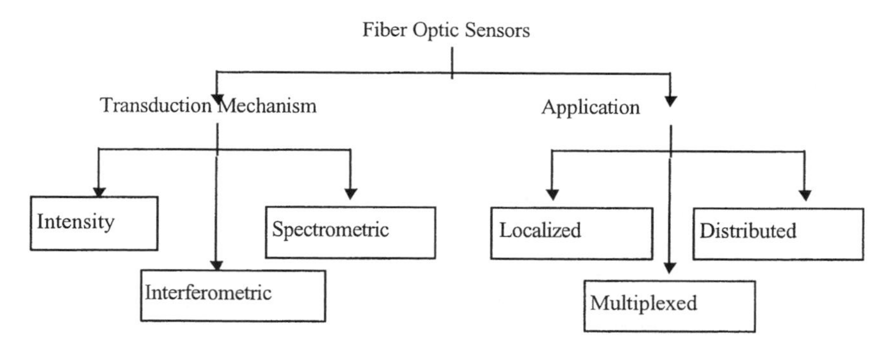

1. SENSOR TYPES

Localized fiber optic sensors determine the measurand over a specific segment of the optical fiber, and are similar in that sense to conventional strain or temperature gauges.

Sensing based on intensity modulation pertains to light intensity losses that are associated with straining of optical fibers along any portion of their length. Sensors taking advantage of this phenomenon are termed as the Intensity or Amplitude type sensor. Spectrometric sensors are widely employed in sensing of chemical reactions, and remote monitoring of contaminants in ground water (Hirschfeld et al. 1983). The transduction mechanism in these types of sensors is based on relating the changes in the wavelength of light to the measurand of interest, i.e. strain. An example of such sensors for measuring strains are Bragg grating type fibers (Morey et al. 1989) (Fig.1).

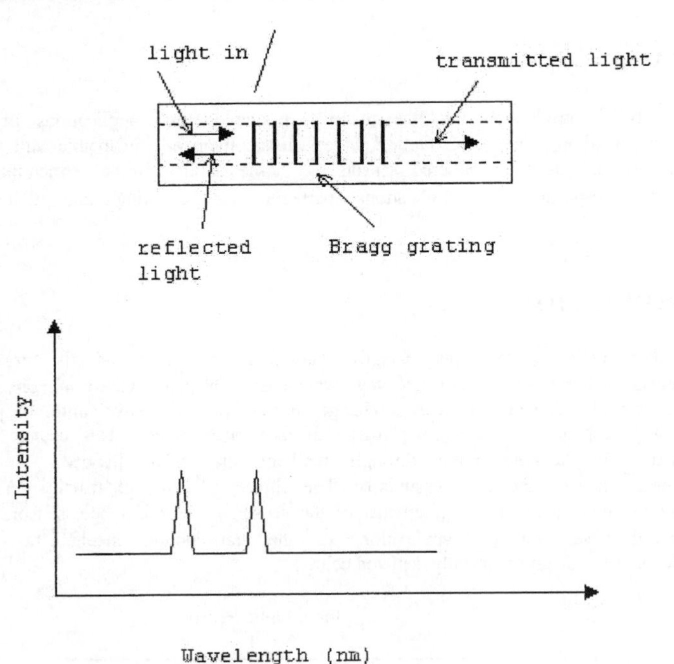

Fig.1 Strain induced shift in wavelength for a fiber optic Bragg grating.

Phase sensors cover a broad range of optical phenomena for sensing purposes. A number of different configurations can be employed for measuring the change in the phase of light by an interferometric sensor (Fig.2). Interferometric sensors are highly sensitive for measuring strains. However, they require interference of light from two identically similar single mode fibers, one of which is used as a reference arm and the other is the actual sensor. The change in phase of the light wave due to strain can be expressed by:

$$\Delta\phi = \varepsilon \cdot l \cdot (\frac{2\pi}{\lambda})(\eta + 1) \tag{1}$$

where,

$\Delta\phi$ = phase shift of the lightwave due to strain

λ = wavelength of the lightwave

ε = applied strain

l=length of the fiber

η = an optical parameter which is a function of the refractive index

An exception to a two arm interferometric sensor is a single fiber Fabry-Perot type sensor (Claus et al. 1993). In a Fabry perot type sensor, the fiber is manipulated in such a way so as to form two parallel reflectors (mirrors), perpendicular to the axis of the fiber. The interference of the reflected signals which are formed in the cavity by the two partial mirrors create the interference pattern. Fabry-Perot sensor is only capable of providing localized measurements at the cavity formed by the two mirrors. The interference pattern generated at the output end of phase sensors is sinusoidal in shape and is directly related to the intensity of applied strain field.

2. MULTIPLEXED AND DISTRIBUTED SENSORS

Multiplexed sensors are usually constructed by combining a number of individual sensors for measurement of perturbations over a large structure. Theoretically, it is possible to use optical switching and other innovative ideas for this purpose. A promising technique is based on wavelength division multiplexing by using Bragg gratings (Kersey et al. 1993). In using this technique, a broad band light source defined as light containing a number of wavelengths within a certain region of spectrum is employed for scanning a number of Bragg grating type sensors in series and/or in parallel. The reflectance wavelength of each Bragg is slightly different from the other. In this way, wavelength shifts of individual sensors are recognized, detected and then related to the magnitude of strain at specific sensor locations.

Distributed sensors make full use of optical fibers, in that each element of the optical fiber is used for both measurement and data transmission purposes. The purpose for making measurements by distributed or multiplexed optical fibers is to determine locations and values of measurands along the entire length of fiber. These sensors are most appropriate for large structure applications, due to their multi-point measurement capabilities. A distributed sensor permits measurement of a desired parameter as a function of length along the fiber. The most widely employed distributed sensing technique is based on measurement of propagation time delays of light traveling in the fiber based on the measurand-induced change in the transmission of light. An optical time domain reflectometer (OTDR) is used for this purpose (Tateda et al. 1989). A pulsed light signal is transmitted into one end of the fiber, and light signals reflected from a number of partial reflectors along the fiber length are recovered from the same fiber end. By using this concept, it is possible to determine the distance to the strain field, d, by way of the two-way propagation time delay, 2t, through the simple relationship (relating velocity and distance):

$$d = 2t.v \qquad (2)$$

where v is velocity of light in the fiber, and $2t$ is the time required for the two-way travel of the signal from individual reflectors. Since, the velocity of light is known, OTDR is capable of detecting the location of strain fields (d_1 ...d_n) through measurement of reflected time signals (Fig. 3).

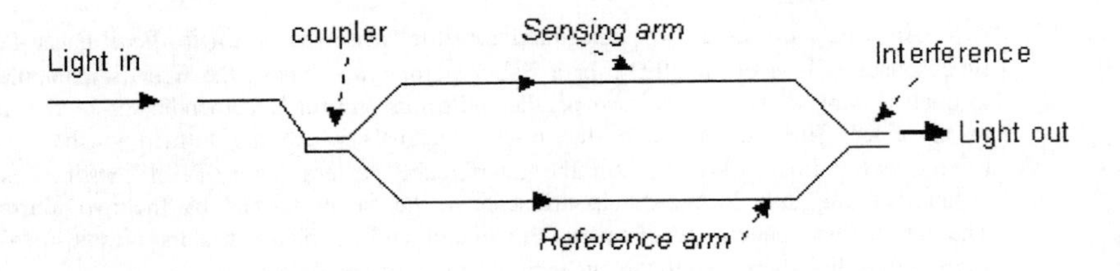

(a) Mach Zender

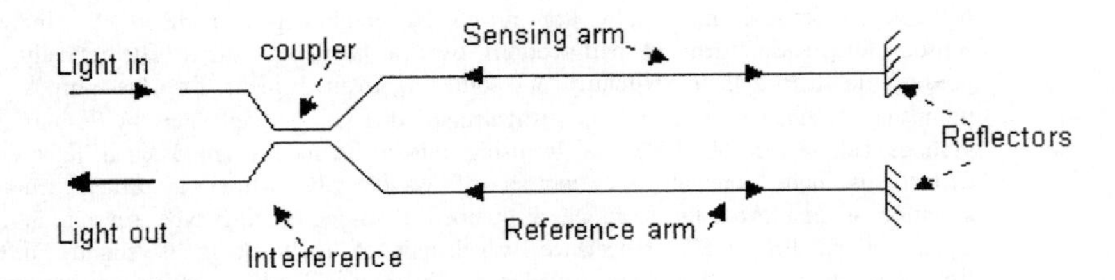

(b) Michelson

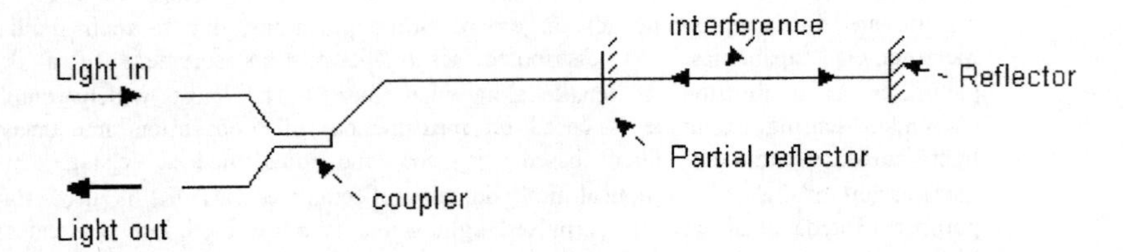

(c) Fabry Perot

Fig.2. Fiber optic interferometric sensors.

3. APPLICATIONS

Some of the Civil Engineering related applications of fiber optic sensors are briefly presented here. Ansari (1990) developed a fiber optic sensor for determination of air content in freshly mixed concrete. The sensor was based on the measurement of the reflected intensity of light through the tip of an optical fiber in contact with fresh concrete. As the sensor tip moves through fresh concrete, it comes in contact with concrete constituents including aggregates, paste, water, and air. Due to the large difference between the refractive index of the fiber ($n=1.46$) and the air bubble ($n=1.0$) at the fiber-air bubble interface, most of the optical intensity will be reflected back in the optical fiber. The kinematics of relative motion between the sensor tip and individual constituents of varying dimensions in concrete will give rise to a reflectivity pattern indicative of percent air in concrete.

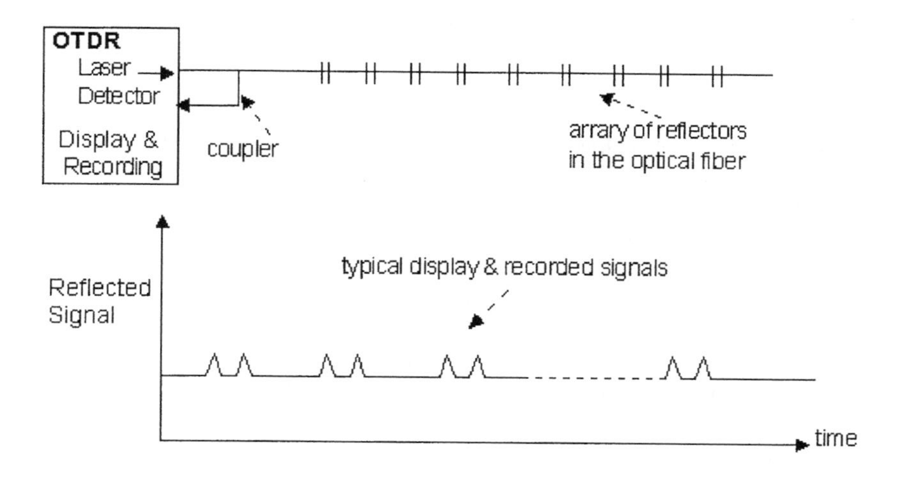

Fig.3 Distributed sensing by way of an OTDR

Rossi, et.al.(1989) embedded a series of optical fibers in laboratory concrete beams as well as in concrete caissons for larger structures. His research involved crack detection due to light intensity loss in the embedded optical fiber. The principle is based on the finding that an optical fiber embedded in a piece of concrete breaks as soon as a crack propagating in the material reaches the fiber causing complete disappearance of the luminous signal transmitted through the fiber. Fuhr et. al.(1992) reports the installation of fiber optic sensors into the Stafford building at the University of Vermont. This study describes the practical issues involved for successful embedment of fiber optic sensors in large concrete structures. These include the construction crew, embedment techniques, and development of schemes for protection of optical fibers when forms are removed from the concrete elements. Wolff et. al.(1992) employed fiber optic sensors for monitoring of prestressing force, and crack formations in Schiessbergstrasse triple span (total span length of 53.0 meters) in Germany. The three-span concrete slab bridge is designed with partial prestressing comprising 27 glass fiber prestressing tendons. For measurement of prestressing force, Wolff et. al. integrated four of the optical fiber sensors within the tendon during its fabrication. Wolff reports crack detection capabilities with an accuracy of 0.15 mm. Masri et al.(1994) employed Fabry-Perot sensors in a one-third scale model of a reinforced concrete multistory frame joint prototype. His research reports successful measurement of strains under cyclic loading conditions. Habel et al.(1994) also reports on using Fabry-Perot sensors for condition monitoring of cracked box girder in Berlin's Charlottenburg bridge.

4. CONCLUSIONS

Fiber optic sensor principles have been briefly described in this article. A number of investigators demonstrated that the integrated sensing systems could also monitor the health of various structural components leading to improved quality control.

REFERENCES

Ansari,F. (1990), "A New Method for Assessment of Air Voids in Plastic Concrete." Cement and Concrete Research, 20(6), 901-910.

Fuhr, P.L., Huston, D.R., Kajenski, P.J., and Ambrose, T.P. (1992) ," Performance and Health Monitoring of the Stafford Medical Building Using Embedded Sensors", J. of Smart Materials and Structures, 1, 63-68.

Habel.W.R., and Hofman,D. (1994a), "Determination of Structural Parameters Concerning Load Capacity Based on Fiber Fabry-Perot Interferometers." in 2nd European Conf. On Smart Structures & Materials. Glasgow, 176-79.

Hirschfeld,T., Denton,T., Milanovich,F., and Klainer,S. (1983), "Feasibility of Using Fiber Optics for Monitoring Groundwater Contaminants." Optical Engineering, 22, 527-31.

Masri, S.F., Agbabian, M.S., Abdel Ghaffar, A.M., Higazy, M., Claus,R.O., and deVries, M.J. (1994), "Experimental Study of Embedded Fiber Optic Strain Gauges in Concrete Structures." ASCE-EMD 120, 1696-1717.

Morey, W.W., Meltz, G. and Glenn, D.H. (1989), "Fiber Optic Bragg Grating Sensors," Proc. SPIE. Fiber Optic and Laser Sensors. 1169, 98.

Kersey, A.D., and Morey, W.W. (1993), "Multiplexed Bragg Grating Fiber-Laser Strain Sensor System with Mode-Locked Interrogation.", Electronic Lett. 29, 112.

Rossi, P. and LeMaou, F. (1989), "New Method for Detecting Cracks in Concrete Using Fiber Optics. "RILEM, Materials and Structures. 22 , 437-42.

Tateda, M., and Horiguchi,T. (1989), "Advances in Optical Time Domain Reflectometry." IEEE, J. Lightwave Tech. 7, 1217-1223.

Wolff, R., and Miesseler, H. (1992), "Monitoring of Prestressed Concrete Structures with Optical Fiber Sensors." in Proc. 1st European Conference on Smart Structures and Materials, Glasgow, 23-29.

Fiber Optic Sensors for Long-term Global Structural Monitoring

Daniele Inaudi
SMARTEC SA, Switzerland

MONITORING AS A STRUCTURE MANAGEMENT TOOL

The construction and maintenance of the civil infrastructure represents between 10% and 20% of the public investment in most European countries. In the last decade we have however witnessed an increasing shift from investments in the construction of new structures to the maintenance and the lifetime extension of the existing ones. With the exception of the high-speed train lines, most of the transportation network, including highways and railway, is completed and in service. However, the steady increase of the passengers and goods circulating in the continent, amplified by the free circulation policy introduced by the European Community, is putting the civil infrastructure under a rude test. Many bridges and tunnels built a few tens of years ago need repair and in many case an extension of their bearing capacity and lifetime that exceed the original plans. Besides the direct costs associated with these interventions, the disruption to the normal use of the structures causes additional inconvenience including traffic jams and accidents that carry additional hidden costs.

The authorities managing the civil infrastructures face the challenge of maintaining the transportation network in a satisfactory state using a limited budget and with little perturbation to its normal use. This task is far more complex than that of building new structures and requires new management instruments.

Structural health monitoring is certainly one of the most powerful management tools and is therefore gaining in importance in the civil engineering community. Monitoring is often and mistakenly presented as a security tool. This is however only the case for the few structures that present a high potential danger such as nuclear power plants and dams. For most other structures the security risks are very limited and fortunately we rarely witness casualties due to a structural collapse. For all other structures, monitoring should be seen as a management tool delivering information on the state of a single structure or on a network of structures. In what we call the information age, structural health monitoring closes the gap between the seemingly inert world of structures and the frenetic one of information technology.

A typical health monitoring system is composed of a network of sensors that measure the parameters relevant to the state of the structure and its environment. For civil structures such as bridges, tunnels, dams, geostructures, power plants, high-rise buildings and historical monuments, the most relevant parameters are:

- Physical quantities: position, deformations, inclinations, strains, forces, pressures, accelerations, vibrations.

- Temperatures.
- Chemical quantities: humidity, pH, chlorine concentration.
- Environmental parameters: air temperature, wind speed and direction, irradiation, precipitation, snow accumulation, water levels and flow, pollutant concentration.

Conventional sensors based on mechanical and/or electrical transducers are able to measure most of these parameters. In the last few years, fiber optic sensors have made a slow but significant entrance in the sensor panorama. After an initial euphoric phase when optical fiber sensors seemed on the verge of invading the whole world of sensing, it now appears that this technology is only attractive in the cases where it offers superior performance compared to the more proven conventional sensors. The additional value can include an improved quality of the measurements, a better reliability, the possibility of replacing manual readings and operator judgement with automatic measurements, an easier installation and maintenance or a lower lifetime cost. The first successful industrial applications of fiber optic sensors to civil structural monitoring demonstrate that this technology is now sufficiently mature for a routine use and that it can compete as a peer with conventional instrumentation.

STRAIN VS. LONG-GAGE SENSORS

The monitoring of a structure can be approached either from the material or from the structural point of view. In the first case, monitoring will concentrate on the local properties of the materials used in the construction (e.g. concrete, steel, and timber) and observe their behavior under load or aging. Short base-length strain sensors are the ideal transducers for this type of monitoring approach. If a very large number of these sensors are installed at different points in the structure, it is possible to extrapolate information about the behavior of the whole structure from these local measurements. Since it is impossible to cover the whole structure with such sensors, some a-priori knowledge about the most interesting and representative locations to be analyzed is required. It will than be assumed that the rest of the structure lying between the measurement points will behave in a similar or predictable way.

In the structural approach, the structure is observed from a geometrical point of view. By using long-gage sensors with measurement bases of the same order of magnitude as the typical size of the structure (for example a few meters for a bridge), it is possible to gain information about the deformations of the structure as a whole and extrapolate on the global behavior of the construction materials. The structural monitoring approach will detect material degradation like cracking or flow only if they have an impact on the form of the structure. Since it is reasonably possible to cover the whole structure with such sensors, no a-priori knowledge about the position of the degradations to be observed (for example the exact position of a crack) is required. This approach usually requires a reduced number of sensors when compared to the material monitoring approach.

The availability of reliable strain sensors like resistance strain gages or, more recently, of fiber optic strain sensors as Fabry-Perot interferometers and Fiber Bragg Gratings has historically concentrated most research efforts in the direction of material monitoring rather than structural monitoring. This latter has usually been applied using external means like triangulation, dial gages and invar wires. Fiber optic sensors offer an interesting means of implementing structural monitoring with internal or embedded sensors.

FIBER OPTIC SENSOR TYPES

There are a great variety of fiber optic sensors [1, 2, 3, 4] for structural monitoring, developed in both the academic and the industrial areas. Unlike the USA, where most efforts seem concentrated to strain sensing, Europe is developing and producing a great variety of sensors for the most disparate types of measurement and application. In this overview we will concentrate on sensors for civil health monitoring that have reached an industrial level or are at least at the stage of advanced field trials.

The following table resumes the sensor technologies that will be discussed in the next paragraphs:

	Main Measured Parameters	**Selection of active groups and companies (see text for details)**
SOFO	Displacement, long-gauge	SMARTEC, IMAC-EPFL (Switzerland)
Microbending	Displacement, long-gauge	Osmos-DehaCom (France)
Bragg gratings	Strain, temperature, pressure, (displacement)	FOS&S (Belgium) LETI (France) EMPA (Switzerland) Naval Research Laboratory (USA) CiDRA/ Weatherford (USA) Blue Road Research (USA) and many more...
Fabry-Perot	Strain	Fiso, Roctest (Canada) Luna Innovations (USA) BAM (Germany)
Raman	Distributed temperature	SENSA/Schlumberger (UK/USA)
Brillouin	Distributed temperature and strain	Omnisens, MET-EPFL, SMARTEC (Switzerland) ANDO (Japan)
Hydrogel	Humidity, water ingress	Univ. of Strathclyde, Glasgow (UK)

Table 1 Selection of Fiber Optic Sensors for civil structural monitoring.

SOFO DISPLACEMENT SENSORS

The SOFO system is a fiber optic displacement sensor with a resolution in the micrometer range and an excellent long-term stability. It was developed at the Swiss Federal Institute of Technology in Lausanne (EPFL) and is now commercialized by SMARTEC in Switzerland [5, 6, 7].

The sensor consists of a pair of optical fibers installed in the structure to be monitored. One of the fibers, called measurement fiber, is in mechanical contact with the host structure itself, being attached to it at the two anchorage points, pre-tensioned in between and protected with a plastic pipe. The other, the reference fiber, is placed loose in the same pipe. All deformations of the structure will then result in a change of the length difference between these two fibers. To make an absolute measurement of this path unbalance, a double Michelson interferometer in is used. The first interferometer is made of the measurement and reference fibers, while the second is contained in the portable reading unit. This second interferometer can introduce, by means of a scanning mirror, a well-known path unbalance between its two arms. Because of the reduced coherence of the source used (the 1.3 micron radiation of a LED), interference fringes are detectable only when the reading interferometer exactly compensates the length difference between the fibers in the structure. If this measurement is repeated at successive times, the evolution of the deformations in the structure can be followed in time.

The precision and stability obtained by this setup have been quantified in laboratory and field tests to 2 micron (2/1000 mm), independently from the sensor length over more than five years. Even a change in the temperature of in fiber transmission properties does not affect the precision, since the displacement information is encoded in the coherence of the light and not in its intensity.

More than 2600 SOFO sensors have been successfully used to monitor more than 100 structures, including bridges, tunnels, piles, anchored walls, dams, historical monuments, nuclear power plants as well as laboratory models.

MICROBEND DISPLACEMENT SENSORS

An alternative fiber optic sensor useful for the measurement of length variations is based on the principle of microbending. In that setup, an optical fiber is twisted with one ore more other fibers or with metallic wires [8] along its sensing length. When this fiber optic twisted pair is elongated, the fibers will induce bending in one-another and cause part of the light to escape from the fiber. By measuring the intensity of the transmitted light it is therefore possible to reconstruct the deformation undergone by the structure on which the sensor is mounted.

A system based on this principle has been marketed for some years through Sicom and more recently by Osmos Deha-Com in France. Typically obtainable resolutions are of 30 μm for short periods (below one day) and 100 μm for the long-term. Microbending sensors are conceptually simple, however temperature compensation, intensity drifts, system calibration and the inherently non-linear relationship between intensity and elongation still present some challenges. This type of sensor seems particularly appropriate for short-term and dynamic monitoring as well as for issuing alarms.

BRAGG GRATING STRAIN SENSORS

Bragg gratings are periodic alterations in the index of refraction of the fiber core that can be produced by adequately exposing the fiber to intense UV light. The produced gratings typically have length of the order of 10 mm. If white light is injected in the fiber containing the grating, the wavelength corresponding to the grating pitch will be reflected while all other wavelengths will pass through the grating undisturbed. Since the grating period is strain and temperature dependent, it becomes possible to measure these two parameters by analyzing the spectrum of the reflected light [9]. This is typically done using a tunable filter (such as a Fabry-Perot cavity) or a spectrometer. Resolutions of the order of 1 $\mu\varepsilon$ and 0.1 °C can be achieved with the best demodulators. If strain and temperature variations are expected simultaneously, it is necessary to use a free reference grating that measures the temperature alone and use its reading to correct the strain values. Setups allowing the simultaneous measurement of strain and temperature have been proposed, but have yet to prove their reliability in field conditions. The main interest in using Bragg gratings resides in their multiplexing potential. Many gratings can be written in the same fiber at different locations and tuned to reflect at different wavelengths. This allows the measurement of strain at different places along a fiber using a single cable. Typically, 4 to 16 gratings can be measured on a single fiber line. It has to be noticed that since the gratings have to share the spectrum of the source used to illuminate them, there is a trade-off between the number of grating and the dynamic range of the measurements on each of them.

Because of their length, fiber Bragg gratings can be used as replacement of conventional strain gages and installed by gluing them on metals and other smooth surfaces [10]. With adequate packaging they can also be used to measure strains in concrete over basis length of typically 100 mm.

A large number of research and development projects for this type of sensors are underway worldwide [11]. In north America, many research and industrial projects are underway to develop monitoring systems based on Bragg Gratings. Just to name a few, the US Naval Research Laboratory has an extensive monitoring project, CiDRA is developing a large verity of fiber optic sensors for the oil and gas intrustry, Blue Road Research is developing multi-axis sensors and insturmenting refurbished bridges in Oregon and Elecrophotonics in Canada produces Bragg-based monitoring systems. Two European projects (STABILOS [12] and COSMUS) focussd on the application of this technology to the measurement of movements in tunnels, mines and other geostructures. In particular, an array of Bragg grating has been installed in the Mont Terri tunnel in Switzerland. The LETI group in France has also used this technology to monitored lock gates [13] and is introducing the system in the nuclear power industry [14], while EMPA (Swiss Federal Laboratories for Materials Testing and Research) has installed them in the Luzzone Dam [15] and in a cable-stayed bridge. Finally, the University of Cantabria in Spain is developing sensors for the electrical power generation industry including strain and acceleration sensors (also base on other sensing techniques) [16]. A comprehensive review by Pierre Ferdinand on the applications of Bragg gratings in Europe can be found in the references [11]

FABRY-PEROT STRAIN SENSORS

Extrinsic Fabry-Perot Interferometers (EFPIs) are constituted by a capillary silica tube containing two cleaved optical fibers facing each others, but leaving an air gap of a few microns or tens of microns between them. When light is launched into one of the fibers, a back-reflected interference signal is obtained. This is due to the reflection of the incoming light on the glass-to-air and on air-to-glass interfaces. This interference can be demodulated using coherent or low-coherence techniques to reconstruct the changes in the fiber spacing. Since the two fibers are attached to the capillary tube near its two extremities (with a typical spacing of 10 mm), the gap change will correspond to the average strain variation between the two attachment points [1,9].

In north America, companies like Sensa in the USA and FISO in Canada offer interesting sensors based on this technology. Contrary to the rest of the world, Europe seems to pay relatively little attention to this interesting sensor technique. A notable exception is the group at BAM in Berlin (Germany), that is using these sensors to monitor the early-age deformations of mortars [17] and has applied them to the monitoring of a concrete bridge in Charlottenbourg [18].

RAMAN DISTRIBUTED SENSORS

Raman scattering is the result of a non-linear interaction between the light traveling in a fiber and silica. When an intense light signal is shined into the fiber, two frequency-shifted components called respectively Raman Stokes and Raman anti-Stokes, will appear in the back-scattered spectrum. The relative intensity of these two components depends on the local temperature of the fiber. If the light signal is pulsed and the back-scattered intensity is recorded as a function of the round-trip time, it becomes possible to obtain a temperature profile along the fiber [19]. A system based on Raman scattering is commercialized by Sensa-Schlumberger in the UK/USA. Typically a temperature resolution of the order of 1°C and a spatial resolution of less than 1m over a measurement range up to 10 km is obtained for multi-mode fibers. A new system based on the use of singlemode fibers should extend the range to about 30km with a spatial resolution of 8 m and a temperature resolution of 2°C.

BRILLOUIN DISTRIBUTED SENSORS

Brillouin scattering sensors show an interesting potential for distributed strain and temperature monitoring [20]. Systems able to measure strain or temperature variations of fibers with length up to 50 km with spatial resolution down in the meter range are now demonstrating their potential in the first field trials. For temperature measurements, the Brillouin sensor is a strong competitor to systems based on Raman scattering, while for strain measurements it has practically no rivals.

Brillouin scattering is the result of the interaction between optical and sound waves in optical fibers. Thermally excited acoustic waves (phonons) produce a periodic modulation of the refractive index. Brillouin scattering occurs when light propagating in the fiber is diffracted backward by this moving grating, giving rise to a frequency shifted component by a phenomenon similar to the Doppler shift. This process is called spontaneous Brillouin scattering.

Acoustic waves can also be generated by injecting in the fiber two counter-propagating waves with a frequency difference equal to the Brillouin shift. Through electrostriction, these two waves will give rise to a travelling acoustic wave that reinforces the phonon population. This process is called stimulated Brillouin amplification. If the probe signal consists in a short light pulse and its reflected intensity is plotted against its time of flight and frequency shift, it will be possible to obtain a profile of the Brillouin shift along the fiber length.

The most interesting aspect of Brillouin scattering for sensing applications resides in the temperature and strain dependence of the Brillouin shift [21]. This is the result of the change the acoustic velocity according to variation in the silica density. The measurement of the Brillouin shift can be approached using spontaneous or stimulated scattering. The main challenge in using spontaneous Brillouin scattering for sensing applications resides in the extremely low level of the detected signal. This requires sophisticated signal processing and relatively long integration times. A commercial system based on spontaneous Brillouin scattering is available from ANDO (Japan).

Systems based on the stimulated Brillouin amplification have the advantage of working with a relatively stronger signal but face another challenge. To produce a meaningful signal the two counter-propagating waves must maintain an extremely stable frequency difference. This usually requires the synchronization of two laser sources that must inject the two signals at the opposite ends of the fiber under test. The MET (Metrology laboratory) group at Swiss Federal Institute of Technology in Lausanne (EPFL) proposed a more elegant approach [22]. It consists in generating both waves from a single laser source using an integrated optics modulator. This arrangement offers the advantage of eliminating the need for two lasers and intrinsically insures that the frequency difference remains stable independently from the laser drift. Omnisens and SMARTEC (Switzerland) commercialize a system based on this setup and named DiTeSt. It features a measurement range of 10 km with a spatial resolution of 1 m or a range of 25 km with a resolution of 5 m. The strain resolution is 20 $\mu\varepsilon$ and the temperature resolution 1°C. The system is portable and can be used for field applications. These values are close to the theoretical limits for a Brillouin system.

Since the Brillouin frequency shift depends on both the local strain and temperature of the fiber, the sensor setup will determine the actual sensitivity of the system. For measuring temperatures it is sufficient to use a standard telecommunication cable. These cables are designed to shield the optical fibers from an elongation of the cable. The fiber will therefore remain in its unstrained state and the frequency shifts can be unambiguously assigned to temperature variations. If the frequency shift of the fiber is known at a reference temperature it will be possible to calculate the absolute temperature at any point along the fiber. Measuring distributed strains requires a specially designed sensor. A mechanical coupling between the sensor and the host structure along the whole length of the fiber has to be guaranteed. To resolve the cross-sensitivity to temperature variations, it is also necessary to install a reference fiber along the strain sensor. Similarly to the temperature case, knowing the frequency shift of the unstrained fiber will allow an absolute strain measurement.

HYDROGEL DISTRIBUTED HUMIDITY SENSORS

Many of the degradations that can occur to the most used structural materials: concrete and steel, have a chemical origin. It is therefore interesting to monitor the presence and the concentration of potentially harmful chemicals such as humidity, chlorine as well as the variations of pH. Chemical measurements with fiber optic sensors are much less developed than those of physical parameters and temperature. It is therefore interesting to cite the development of a distributed humidity sensor that is based on the use of a particular hydrogel capable of transforming a humidity variation in a change in its dimensions [23]. This allows the transformation of a difficult chemical measurement in a much easier strain or elongation measurement. A first sensor, developed at Strathclyde University, is based on a hydrogel that swells when wetted. The expansion of the hydrogel induces microbending losses in an optical fibers that can be detected with a standard Optical Time Domain Interferometer (OTDR). The system shows potential for measurement of water ingress and humidity in large structures and in areas that are difficult to inspect. In one of the first field demonstrations, the system was used to detect an incomplete grouting of a post-tensioning cable duct [24]. By using another type of hydrogel, it is expected that this type of sensors will be capable of measuring other chemicals and in particular the pH chances associated with carbonation in concrete.

OUTLOOK

In the first decade of Fiber Optic Sensing technology, most efforts were concentrated on the different subsystems. The demodulators and the multiplexing architectures have seen important developments and many technologies are today mature for field and industrial applications. Some techniques have emerged like fiber Bragg grating sensors, low-coherence sensors and external Fabry-Perot interferometers, others are living a second youth like intensity based sensors. New technologies, like Brillouin scattering, are still in the development phase and many more will certainly emerge in the future. Portable reading units are getting smaller each year and have been successfully operated in demanding environment like those found in marine and civil engineering applications.

In these last few years, the maturity of the reading unit subsystems has driven toward the development of reliable sensors and installation techniques. Fiber optic sensors have been embedded successfully in a number of materials and structures including composites, concrete, timber and metals. Some of these efforts are leading to industrial products and this will allow the instrumentation of structures with an increasing number of sensor at reasonable prices. It can be estimated that a few thousands fiber optic sensors have been installed to date in civil structures alone. The improvement of packaging techniques and the reduction of costs will also be helped by the continuous development of fiber optic components like fibers, cables, connectors, couplers and optical switches driven by the much larger telecommunication market.

With structures equipped with hundreds or even thousands of sensors, measuring different parameters each second, the need of automatic data analysis tools will become increasingly urgent. Efforts have already been directed in this direction. Unfortunately, each type of structure and sensor needs specific processing algorithms. Vibration and modal analysis have attracted many research efforts and geometrical analysis like curvature measurements can be easily applied to different types of structures like bridges, tunnels or spatial structures. Many other concepts like neural networks, fuzzy logic, artificial intelligence, genetic algorithms and data mining tools will certainly find an increasing interest for smart processing applications.

The ubiquity of digital networks and cellular communication tools increases the flexibility of the interfacing and makes remote sensing not only possible but even economically attractive. Of course every remote sensing system has to be based on reliable components since the need of manual interventions obviously reduces the interest of such systems.

Smart structures will both demand and produce sophisticated smart sensing and processing systems. Continuous developments in actuators based on piezoelectric materials and shape memory alloys complement ideally the progress made in sensor and processing technology. Most efforts are directed towards vibration damping, noise reduction and shape control, mainly for the aeronautics and space industry. Civil engineering is also producing interesting smart structures applications in particular for seismic control and many experiments have been conducted at least on reduced scale models. Other applications like vibration and modal control of large civil structures like suspended bridges could be potentially interesting but the forces required to achieve these results are still exceedingly high. In a first phase we can expect that smart structures will be used to increase the comfort of the users and the life-span of the structures by reducing the amplitude of its oscillations under seismic, traffic or aerodynamic loads. These systems will not have a major structural role and their failure would not lead to important structural damages. The acceptance of smart structures where the control system plays a structural role will require well-proved and reliable systems and will probably appear first in high-risk structures like fighters airplanes or space structures.

More than the developments in each of the smart sensing subsystems, it is however the successful integration of different technologies that will lead to increasingly useful applications. This integration is possible only in highly multidisciplinary teams including structural, material and sensor engineers. The necessary competencies already exist in many industries and universities but have to be brought together and adapted to each other needs.

The final judge of Fiber Optic Sensors systems will however be the market. Even well designed and perfectly functioning systems will have to prove their economic interest in order to succeed. Unfortunately the evaluation of the benefits of a sensing system is often difficult and the initial additional investments are paid back only in the long run. Furthermore it is not easy to quantify the benefits of the increased security of one structure or of a better knowledge of its aging characteristics. In many fields including civil engineering and aeronautics we are however witnessing an investment shift from the construction of new structures to the maintenance and the life-span extension of the existing ones. In these domains, smart sensing technologies have certainly an important role to play.

CONCLUSIONS

The monitoring of new and existing structures is one of the essential tools for a modern and efficient management of the infrastructure network. Sensors are the first building block in the monitoring chain and are responsible for the accuracy and reliability of the data. Progress in the sensing technology can therefore be produced by more accurate measurements, but also from systems that are easier to install, use and maintain. In the recent years, fiber optic sensors have moved the first steps in structural monitoring and in particular in civil engineering. Different sensing technologies have emerged and quite a few have evolved into commercial products.

It is difficult to find a common reason for the success of so diverse types of sensors, each one seems to have found a niche where it can offer performance that surpass or complement the ones of the more traditional sensors. If three characteristics of fiber optic sensors should be highlighted as the probable reason of their present and future success, I would cite the stability of the measurements, the potential long-term reliability of the fibers and the possibility of performing distributed and remote measurements. In the near future it is therefore to expect that fiber optic sensors will consolidate their presence in the structural sensing industry.

REFERENCES

[1] Udd, E., *Fiber Optic Sensors*, Wiley, 1991.

[2] Udd, E., *Fiber optic smart structures*, Wiley, New York, 1995

[3] D. Inaudi, “Fiber optic smart sensing” *Optical Measurement techniques and applications*, P. K. Rastogi editor, Artech House, pp. 255-275, 1997

[4] Proceedings of the Optical Fiber Sensor (OFS) series. The latest being: "12th International Conference on Optical Fiober Sensors", Williamsbourgh USA, October 1997, OSA 1997 Technical Digest Series Vol 16.

[5] D. Inaudi, A. Elamari, L. Pflug, N. Gisin, J. Breguet, S. Vurpillot "Low-coherence deformation sensors for the monitoring of civil-engineering structures", *Sensor and Actuators A*, Vol. 44, pp. 125-130, 1994

[6] D. Inaudi, "Field testing and application of fiber optic displacement sensors in civil structures", 12th International conference on OFS '97- Optical Fiber Sensors, Williamsbourg, OSA Technical Digest Series, Vol. 16, pp. 596-599, 1997

[7] D. Inaudi, N. Casanova, P. Kronenberg, S. Marazzi, S. Vurpillot, "Embedded and surface mounted fiber optic sensors for civil structural monitoring", Smart Structures and Materials Conference, San Diego, SPIE Volume 3044, pp. 236-243, 1997

[8] L.Falco, O. Parriaux, "Structural metal coatings for distributed fiber sensors", Opt. Fiber Sens. Conf. Proc., pp. 254, 1992

[9] A. Kersey, "Optical Fiber Sensors" *Optical Measurement techniques and applications*, P. K. Rastogi editor, Artech House, pp. 217-254, 1997

[10] S. T. Vohra, B. Althouse, Gregg Johnson, S. Vurpillot and D. Inaudi, "Quasi-Static Strain Monitoring During the 'Push' Phase of a Box-Girder Bridge Using Fiber Bragg Grating Sensors", European Workshop on Optical Fibre Sensors, Peebls Hydro, Scotland, July 1998

[11] P. Ferdinand et al. "Application of Bragg grating sensors in Europe", 12th International conference on OFS '97- Optical Fiber Sensors, Williamsbourg, OSA Technical Digest Series, Vol. 16, pp. 14-19, 1997.

[12] P. Ferdinand et al. "Mine Operating Accurate Stability Control with Optical Fiber Sensing and Bragg Grating technology: the Brite-EURAM STABILOS Project", OFS 10 Glasgow, 1994 pp. 162-166. Extended paper: Journal of Lightwave Technology, Vol. 13, No. 7, pp. 1303-1313, 1995

[13] M. Bugaud, P. Ferdinand, S. Rougeault, V. Dewynter-Marty, P. Parneix, D. Lucas, "Health Monitoring of Composite Plastic Waterworks Lock Gates using in-Fiber Bragg Grating Sensors", 4t^h European Conference on Smart Structures and Materials, July 1998, Harrogate, UK

[14] P. Ferdinand et al. "Potential Applications for Optical Fiber Sensors and Networks within the Nuclear Power Industry" in *Optical Sensors*, J. M. Lopez-Higuera ed., Universidad de Cantabria

[15] R. Brönnimann, Ph. Nellen, P. Anderegg, U. Sennhauser "Packaging of Fiber Optic Sensors for Civil Engineering Applications", Symposium DD, Reliability of Photonics Materials and Structures, San Francisco, 1998, paper DD7.2

[16] J. M. Lopez-Higuera, M. Morante, A. Cobo "Simple Low-frequency Optical Fiber Accelerometer with Large Rotating Machine Monitoring Applications", Journal of Lightwave Technology, Vol. 15, No. 7, pp. 1120-1130, July 1997

[17] W. Habel et al. "Non-reactive Measurement of Mortar Deformation at Very Early Ages by Means of Embedded Compliant Fiber-optic Micro Strain Gages", 12th Engineering Mechanics ASCE Conference, La Jolla USA, May 1998

[18] W. Habel, D. Hofmann "Determination of Structural Parameters Concerning Load Capacity Based on Fiber Fabry-Perot-Interferometers", Proc. SPIE, Vol. 2361, 1994, pp. 176-179

[19] Dakin, J. P. et al., "Distributed optical fiber Raman temperature sensor using a semiconductor light source and detector", Proc, IEE Colloq. on Distributed Optical Fiber sensors, 1986

[20] T. Karashima, T. Horiguchi, M. Tateda, "Distributed Temperature sensing using stimulated Brillouin Scattering in Optical Silica Fibers", *Optics Letters*, Vol. 15, pp. 1038, 1990

[21] M. Niklès, L. Thévenaz, P. Robert, "Brillouin Gain Spectrum Characterization in Single-Mode Optical Fibers", *Journal of Lightwave Technology*, Vol. 15, No. 10, pp. 1842-1851, October 1997

[22] M. Niklès et al., "Simple Distributed temperature sensor based on Brillouin gain spectrum analysis", Tenth International Conference on Optical Fiber Sensors OFS 10, Glasgow, UK, SPIE Vol. 2360, pp. 138-141, Oct. 1994

[23] W. C. Michie et al. "A fiber Optic/Hydrogel Probe for distributed Chemical Measurements", OFS 10 Glasgow, 1994 pp. 130-133.

[24] W. C. Michie, I. McKenzie, B. Culshaw, P. Gardiner, A. McGown "Optical Fiber Grout Flow Monitor for Post Tensioned Reinforced Tendon Ducts", Second European Conference on Smart Structures and Materials, Glasgow, October 1994, SPIE Vol. 2361, pp. 186-189.

Section 2

Large-Scale Tests on Smart Structures for their Performance Verification

G. Magonette

ELSA, IPSC, Joint Research Centre, European Commission, 21020 Ispra (Varese), Italy

ABSTRACT

The control of structural vibrations produced by earthquake or wind can be implemented by various methods such as modifying rigidities, damping, or shape, and by providing passive or active counter forces. To date, some methods of structural control based on the use of various active, semi-active, passive and hybrid control schemes have been used successfully and offer great promises. An overview of the experimental activities in structural control performed at the ELSA laboratory is presented. After a short presentation of passive techniques, we illustrate in more detail an innovative way to reduce the vibrations of cable-stayed structures by the application of an active damping strategy. Finally our research program in semi-active control is introduced.

1. INTRODUCTION

An overview of the experimental activities in structural control performed at the ELSA laboratory is presented. Two basic experiments illustrating the capability of the continuous pseudo-dynamic (PsD) method to test structures equipped with strain-rate sensitive isolators and dissipaters are summarized. The concept of testing with substructuring is introduced and the capabilities of this method are underlined. Next testing of an active tendon control system for civil engineering cable-supported structures is illustrated. The active control technique is based on a tendon actuator collocated with a force sensor. A large-scale mock-up using industrial components has been built and extensively analysed. The test-bed is a model of cable-stayed bridge during its construction phase, equipped with hydraulic actuators on the two longest stay-cables. The results of several damping experiences have clearly demonstrated the efficiency of the active damping system. Finally the interest of semi-active control is reminded and the future activities of ELSA laboratory in this field are summarized.

2. PASSIVE CONTROL SYSTEM

A typical problem to face when performing PsD tests on structures fitted with anti-seismic protection devices, which are in many cases constructed with strain-rate sensitive materials, is the compensation of the error affecting the measured restoring forces due to the time-scale expansion inherent to the PsD method. The methodology that has been developed at ELSA and applied to the tests described hereafter consists of performing the PsD test at an affordable speed for the available devices and applying a compensation technique to correct the strain rate effect error. This technique is based on a correction of the measured restoring forces according to a previous characterising test for the specific devices. Such method cannot be applied in a general way because the correction is based on a particular modelling of the strain rate effect that might not be valid for other types of materials. However, it has properly worked for several types of high-damping rubber devices and visco-elastic absorbers that we have used for different projects. The advantages of using this technique are the usual ones of the reduced-speed PsD method which gives the possibility to use large-size specimens with slow testing devices and a high accuracy in the imposition of displacements and measurements of forces. Two specific typologies of passive seismic protection were tested. The first deals with base isolation and the second with energy dissipation devices.

2.1. Seismic Isolation

The concept of introducing an isolation system to reduce the vibration transmitted to floors and neighbouring equipment is well established. The basic elements of a practical system include: a flexible mounting so that the period of the total system is increased sufficiently to reduce the response; a damper or energy dissipaters so that the relative displacement between the structure and the ground can be controlled to a practical level; a method of providing rigidity under low service loads, e.g. wind and minor earthquake.

The first experimental activity conducted in this field aimed to validate the continuous PsD procedure and was conducted on a scaled 5-storey frame structure isolated by means of high damping rubber bearings (HDRBs), which had been tested on the shaking table of ISMES (Italy). The test campaign included essentially two phases. The first action was to characterize the rubber bearing and to develop the strain rate compensation method.

Afterwards, the experimental results obtained by means of the PsD method, operated with strain rate compensation, were compared with those achieved by means of truly dynamic tests on shaking table. The comparative analysis showed that the continuous PsD method could reliably be used to test structures protected with rubber isolation devices (Magonette et al. 1997).

2.2. Passive energy dissipation

The basic role of passive energy dissipation devices when incorporated into a structure is to absorb or consume a portion of the input energy, thereby reducing energy dissipation demand on primary structural members and minimising possible structural damage. Unlike seismic isolation, however, these devices can be effective against wind-excited motions as well as those due to earthquakes. An experimental campaign was conducted on a large-scale reinforced concrete building protected by rubber-based energy dissipation devices (see figures 1 and 2). The choices commonly practiced to meet seismic criteria for reinforced concrete frame structures are mainly based on strengthening of the design.

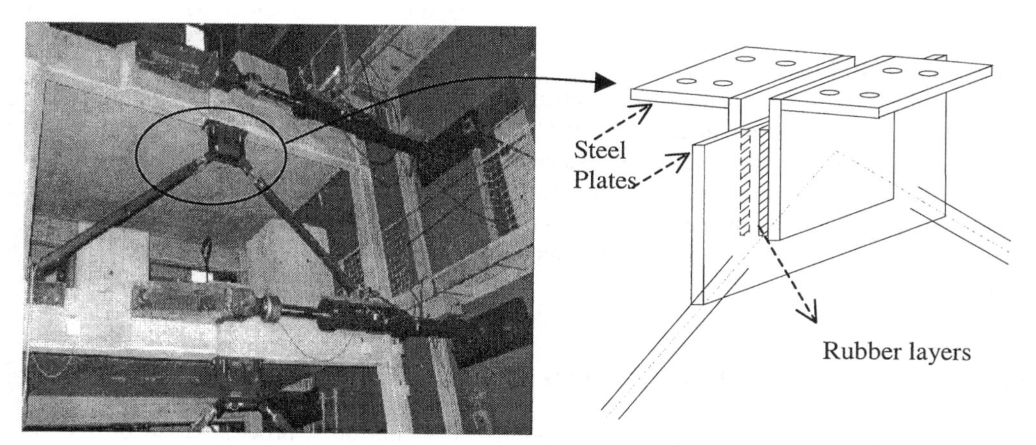

Figure 1. K-bracing and protection devices. PsD actuators with load cell and steel braces

Figure 2. Details of the attachment of the devices to the mock-up structure

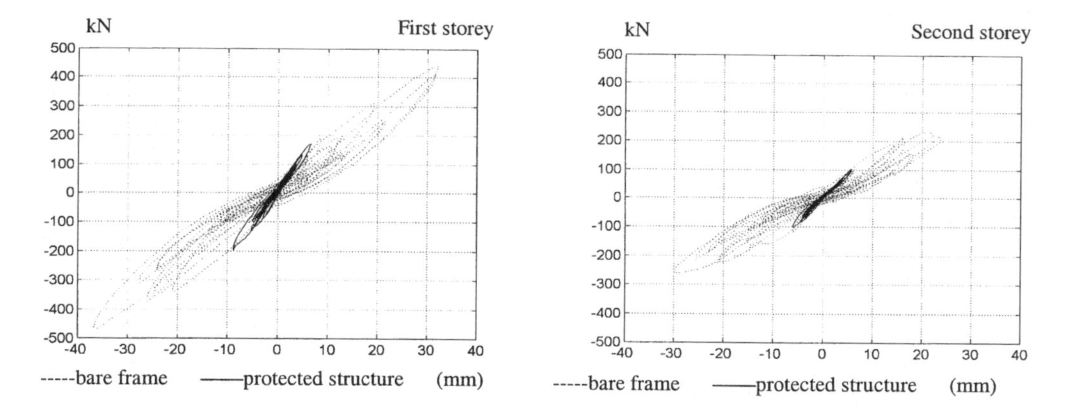

Figure 3. RC Frame shear load versus inter-storey drift, PsD tests of protected structure and bare frame.

The introduction of viscoelastic energy-dissipation devices brings a "soft" alternative to the well-known strengthening method or more recent seismic isolation technology. The experimental tests aimed to verify and quantify the effectiveness of such a system. A two-storey two-bay reinforced concrete specimen of an office building scaled by 2/3 in dimension was designed for PsD testing. The viscoelastic dampers were installed as pairs with the rubber material being deformed in shear. Continuous PsD testing was performed on the protected and on the bare frame using an artificial accelerogram compatible with Eurocode-8 response spectrum. The test was carried out by applying the strain rate compensation correction developed in the seismic isolation tests campaign. Figure 3 demonstrates the effectiveness of the protection devices. Further information can be found in (Taucer et al. 1999). The consistency of the applied correction technique for the compensation of the strain-rate effect was verified by the execution of a small real dynamic random-burst test that was repeated both dynamically and pseudodynamically. In the case of the PsD test with the correction technique the differences with respect to the dynamics test were very small (Molina et al. 2000).

Currently the above-mentioned experimental activity proceeds with seismic testing campaigns on a full-scale three-storey R/C frame structure to be retrofitted by different passive control techniques including hysteretic friction devices, shear panel devices and Jarret visco-elastic absorbers. The rational of this research program is on the whole represented by the experimental verification of possible retrofitting interventions on existing structures by different passive control technologies, within a combined repair-retrofit context.

3. HIGH SPEED CONTINUOUS PsD TESTING ON COMPONENTS.

Considering that the PsD test is in fact a numerically controlled process based on motion equations, one can readily reach the idea that only part of the specimen that has complex hysteretic behaviour may be fabricated and tested in the laboratory, whereas the remaining parts are handled only within the computer. By means of substructuring techniques, the displacements that are imposed on the test structure are obtained by solving the equations of motion of the global system, while the numerical model provides the restoring forces of the portions that are not subjected to experimental testing. With the rapid progress of servo-hydraulic control technologies and numerical integration methods, this concept is becoming nowadays the base for the development of a dynamic testing system in which the critical sections or components of a structure are tested at full or large scale (and possibly in real time) by reproducing the forces and displacements imposed on it by the surrounding structure modelled numerically. This technique is very appealing for testing strain rate sensitive components and semi-active devices and many laboratories are working on its implementation. Because of the very high computing speed needed, it is not always possible to calculate the required displacement increment directly from a non-linear analysis. In such cases an equivalent linearized model is created, which takes as input any applied forces or displacements together with measured forces from the test specimen and will output the required displacements to be imposed on the specimen. At ELSA, to extend the applicability of the continuous PsD method to fast on-line substructuring tests a very significant upgrading of the control systems of the conventional testing facility has been undertaken. The first experimental results obtained are very encouraging and increasing resources are now invested to finalize this testing methodology.

4. ACTIVE CONTROL SYSTEM

An increasing request exists to characterize or verify experimentally active and semi-active devices used for the protection of civil constructions and also to assess the performance of the whole control system operating on large-scale structures under actual loading conditions. Hereafter we introduce recent experiments performed at ELSA on a large-scale model of cable-stayed bridge (ACE Project 1997) protected by active control. The aim of this active system is to upgrade the structural damping of the complete structure and therefore to mitigate the induced vibrations of both the cables and the cable-supported structure. The methodology applied here is based on an active tendon including an actuator collocated with a force sensor. The active damping is driven by the displacement control of the cable anchor point. The control law is the so-called *Integral Force Feedback* (*IFF*). The basic principle of this control is to force the dynamic tension of the cable to produce a work by moving an anchorage according to the variation of the tension measured at the same anchorage. More details are presented in (Aupérin & Dumoulin 2000, Magonette et al. 1999, Preumont 1997). It is widely accepted that the active damping of linear structures is much simplified if one uses collocated actuator-sensor pairs; for non-linear systems, this configuration is still quite attractive, because there exists control laws that are guaranteed to remove energy from the structure. The direct velocity feedback is an example of such "energy absorbing" control. When using a displacement actuator u (active tendon) and a force sensor T, the (positive) *Integral Force Feedback*

$$u = g_p \int T \, dt \tag{1}$$

also belongs to this class, because the power flow from the control system is:

$$W = -T\dot{u} = -g_p T^2 \tag{2}$$

This control law applies to non-linear structures; all the states that are controllable and observable are asymptotically stable for any value of g_p (infinite gain margin). The foregoing theoretical results have been confirmed experimentally with a laboratory scale cable structure.

4.1. Bridge mock-up description

The bridge mock-up is a cable stayed cantilever beam that basically represents a cable-stayed bridge under construction. The deck, about 30 m long, is mainly composed of two H-beams whose axes are spaced 3.0 m apart (figure 4/a). Each H-beam is fixed to the Reaction Wall. The vibration excitation source is anchored at the free end of the deck. Four pairs of parallel stay cables support the deck. The requirements of active control of cable supported structure are to carry both the dynamic loads, i.e. induced by the vibrations of the structure around its nominal position and the static loads due to the dead and live loads. These requirements led to a two stages cylinder architecture: a "static" cylinder compensates the static loads by using an accumulator pressure circuit while a smaller cylinder is in charge of the position control of the piston. The long-term changes of the static loads as well as the temperature variations due to the ambiance were adapted by pressure control of the static load accumulator pressure circuit. (figure 4/b).

4.2. Experimental testing campaign

The mock-up has been exhaustively instrumented with displacement, velocity and acceleration transducers (96 measurement channels were recorded with a sampling rate of 100 Hz). The cable motions were measured with accelerometers and laser scanning systems.

The testing campaign began with modal analysis of the structure. Next, to further confirm the effect of the active damping on the structural response, supplementary tests were conducted using a band-limited white noise excitation. During these experiments, an electro-hydraulic shaker equipped with an inertial mass of 445 kg was mounted at the free edge of the bridge in central position to induce a random excitation. The excitation frequency bandwidth is confined in the range 0.6 Hz to 1,3 Hz. A comparison between two tests executed with and without control is presented in figure 5. In the controlled test we have measured a reduction by a factor 7 of the dynamic component of the tension in the cable. This indicates a substantial reduction of fatigue at the cables anchorage. From the experimental results it has been observed that the whole structure (deck and cables) was strongly damped.

Figure 4. (a) Large scale cable-stayed bridge mock-up
(b) Close view of the hydraulic actuator

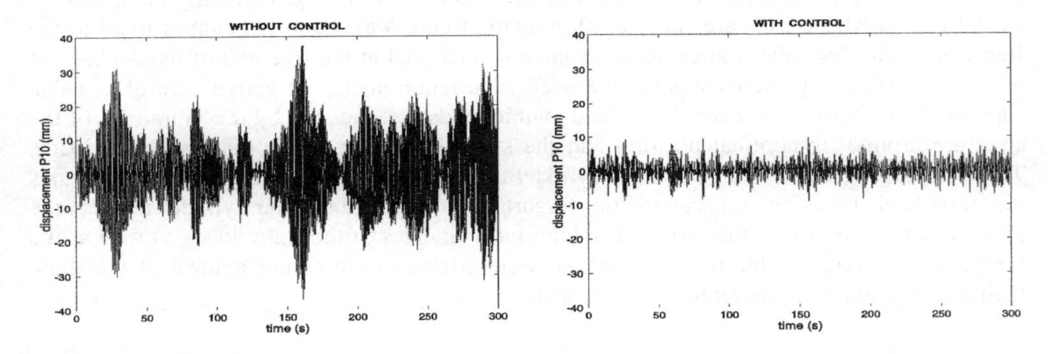

Figure 5. Comparison of the vertical displacement responses at the free end edge of the mock-up in the controlled and uncontrolled tests.

5. FUTURE DEVELOPMENT: SEMI-ACTIVE CONTROL

In the mainframe of the EC-GROWTH program, ELSA is presently involved in a research project called CaSCO (Consistent Semiactive System Control). This research is focused on the size minimization, analysis, experimental evaluation and economic feasibility of semi-active control systems employing advanced material technologies. These innovation products are ideally suited for many applications, including but not limited to the manufacturing sector for vibration and noise isolation. Additional applications are in vibration and noise control of railway lines, increase in energy dissipation of structures, vibration control of tall building under wind load and damping of structures with guy cables.

Semi-active control devices using magnetorheological fluids were selected based on promising preliminary results. Of special interest will be the trade-off between the number and size of the devices. In other words, the dampers will be reduced in size and distributed throughout the structure as a viable means for dynamic hazard mitigation. Distributed installations of many small-scale energy dissipating devices have the advantage to optimise noise and vibration control by locally dissipate dynamic energy before it enters into the global system.

6. CONCLUSION

This paper has presented a short overview of the activity of the ELSA laboratory in structural control. An important part of the research was finalized to the improvement and validation of the continuous PsD testing method for structures protected against earthquake by strain-rate sensitive devices. Furthermore an active control system able to increase largely the structural damping of a complete structure has been tested and validated. The experimental study of semi-active systems is underway and seems very promising.

REFERENCES

ACE Project 1997. Active Control in Civil Engineering, *EC Brite-Euram Contract n°BRPR-CT97-0402.*

CASCO Project 1999. Consistent Semiactive System Control, *EC, GROWTH Programme, Contract N°GIRD-CT-1999-00085.*

Magonette, G., Molina, F. J., Taucer, F., Renda, V. & Tognoli P. 1997, Contribution of the JRC Ispra to the inter-comparison of analysis methods for seismically isolated nuclear structures, *International Post-SMIRT Conference Seminar on Seismic Isolation, Passive Energy Dissipation and Active Control of Vibrations of Structures, Proc. intern. symp.*, Taormina, 25-27 August 1997, Italy.

Magonette G., Marazzi F., Bossens F, & Dumoulin C. 1999, Active Control Experiment of a Large Scale Cable-Stayed Bridge Mock-Up, *Third International Symposium on Cable Dynamics*, Trondheim, 1999, Norway.

Molina, F. J., Verzeletti, G., Magonette, G., and Taucer, F. (2000). Dynamic and Pseudodynamic responses in a two storey building retrofitted with rate-sensitive rubber dissipators. *Proceedings of the 12 World Conference on Earthquake Engineering*. New Zealand.

Preumont A. 1997, *Vibration Control of Active Structures: An Introduction*, Kluwer Academic Publishers.

Taucer, F., Magonette, G. & Marazzi, F. 1999, Seismic Retrofit of a Reinforced Concrete Frame with Energy Dissipation Devices, *Workshop on Seismic Performance of Built Heritage in Small Historic Centres*, April 22-24, 1999, Assisi, Italy.

Vibration Suppression by Energy Pumping

Alexander F. Vakakis[1], D. Michael McFarland[2] and Lawrence A. Bergman[2]

[1]*University of Illinois at Urbana-Champaign and National Technical University of Athens*

[2]*University of Illinois at Urbana-Champaign*

1 INTRODUCTION

The term *energy pumping* refers to the rapid and irreversible transfer of energy from a vibrating mechanical system to an attached *nonlinear energy sink.* This phenomenon can be exploited to quickly reduce the dynamic response of a primary structure subject to direct transient excitation by directing energy to a sink where it can be dissipated without spreading back to the primary substructure. This paper reports on some preliminary numerical and experimental results demonstrating the potential of energy pumping in this application.

2 MECHANICAL MODEL

By combining a single-degree-of-freedom oscillator (the primary system, whose motion is to be reduced) with a nonlinear SDOF system (a nonlinear energy sink, or NES), we obtain the 2-DOF system of Fig. 1. The NES consists of a mass m_1 and a nonlinear spring characterized by the coefficient k_1 and the exponent p_1. For the energy sink to be effective, it is necessary that this spring have no linear term in its force-displacement relation; that is, it should be essentially nonlinear. We assume the nonlinear spring exhibits a cubic hardening restoring force, corresponding to $p_1 = 3$. The sink is furthermore taken to possess linear viscous damping, producing a dissipative force proportional to its velocity with respect to the ground. Coupling of the primary and sink masses is through an undamped spring; a linear spring ($p = 1$) is assumed for simplicity. The coefficient ε of this spring is regarded as a small parameter, reflecting the notion of physically "weak" coupling.

The equations of motion of the system are

$$m\ddot{x} = F - c\dot{x} - kx + \varepsilon(x_1 - x)\,, \tag{1a}$$

$$m_1\ddot{x}_1 = -\varepsilon(x_1 - x) - c_1\dot{x}_1 - k_1 x_1^{p_1}\,. \tag{1b}$$

These can be easily recast as four first-order state equations for simulation purposes, and this state model augmented with auxiliary variables to facilitate the computation of various quantities useful in design work.

Extensive simulations have been carried out in the course of designing a 2-DOF experimental test rig. Typical simulation results are shown in Fig. 2, which depicts the displacement response of a SDOF system alone and when connected to an NES, and in Fig. 3, which shows the associated flow and dissipation of energy in the system with the NES. Noteworthy here is the rapid reduction in the amplitude of vibration of the primary mass due to pumping of energy to the sink compared to the exponentially decaying envelope of the viscously damped SDOF response.

3 LABORATORY TEST RIG

In designing full-scale structures, one typically would want to make the secondary mass a fraction of the primary. For initial simulations of potential experimental hardware, values of $m = 1\text{ kg}$ and $m_1 = 0.5\text{ kg}$ were selected, giving a large but not unrealistic mass ratio of 0.5. It was desired that the natural frequency of the primary structure, which is the dominant frequency of the response, be around 10 Hz so that either accelerometers or LVDTs could be used to measure the motions of the primary and sink masses. Finally, light damping was desirable so that a large portion of the input energy could be clearly shown to flow from the primary structure to the sink before being dissipated.

3.1 Parameter Values Selected through Numerical Simulation

With these considerations in mind, several simulations of the 2-DOF system were carried out by numerically integrating eqs. (1). The initial mass, stiffness and damping values used were similar to those of a simulated system in which energy pumping had previously been noted, but these were iteratively adjusted towards the desired properties described above until the values given in Table 1 were obtained. The transient load was represented as a half-sine pulse of amplitude 80 N and duration 0.02 s.

3.2 Design of the Nonlinear Spring

Several schemes for constructing a nonlinear spring were considered, including some from Rivin (1999), but the need to minimize any linear term in the spring response, the necessity of repeatability of the configuration, and the desire to avoid relying upon potentially frequency-dependent material properties led us to favor an arrangement of linear springs in which a nonlinear force-displacement characteristic derives from geometry alone. The simplest form of this is shown in Fig. 4, which depicts the deformation of a linear spring in response to a load applied perpendicular to its original, undeformed axis.

The force in the spring is proportional to the change in its length,

$$F_s = k(\sqrt{l^2 + x^2} - l)\,, \tag{2}$$

and static equilibrium requires that

$$F = F_s \sin\theta\,. \tag{3}$$

Hence

$$F = k(\sqrt{l^2+x^2} - l)\frac{x}{\sqrt{l^2+x^2}} = kx\left[1 - l(l^2+x^2)^{-1/2}\right]. \tag{4}$$

The Taylor series expansion of the nonlinear term about the point $x = 0$ can be found to be

$$(l^2+x^2)^{-1/2} = \frac{1}{l} - \frac{x^2}{2l^3} + \frac{3x^4}{8l^5} + O(x^6). \tag{5}$$

As usual, we assume x is small in some sense (specifically, with respect to l; see, e.g., Stoker (1950)), so that we may neglect higher-order terms. Retaining terms through second order in x and substituting this series into the expression for F immediately above leads to

$$F \approx kx\left[1 - l\left(\frac{1}{l} - \frac{x^2}{2l^3}\right)\right] = \frac{1}{2l^2}kx^3. \tag{6}$$

One practical realization of this cubic spring is shown in Fig. 5. The linear spring is a piece of piano wire that remains straight under nearly zero tension. The area A, Young's modulus E and half-span l of this wire determine the linear stiffness of each half of the wire in tension,

$$k = \frac{EA}{l}, \tag{7}$$

from which the nonlinear stiffness of the wire in transverse deflection can be computed as

$$F \approx \frac{EA}{l^3}x^3. \tag{8}$$

The nonlinear spring has been calibrated for a number of wire diameters and span lengths by measuring its deflection under a static load. Typical measured data are plotted in Fig. 6 along with the result of fitting a cubic function to these points. The calibration results exhibit the desired nonlinearizable behavior (the slope of the measured and fitted curves approaches zero for small displacements) and show that this design can achieve a large range of nonlinear spring coefficients.

3.3 Two-DOF Test Apparatus

The photo in Fig. 7 shows the apparatus set up in the laboratory. The masses are made of aluminum angle stock, supported by an air track and connected to ground and to one another by rigid rods and linear leaf springs. The cubic spring consists of a piece of piano wire supported above and perpendicular to the air track, and is attached at its center to the sink mass.

The linear portions of the test structure have been characterized by experimental modal analysis. Table 2 summarizes the physical parameters of the system.

4 EXPERIMENTAL ENERGY PUMPING RESULTS

The following figures show the measured accelerations of the primary and sink masses of the system described above.

The data plotted in Fig. 8 were taken while the nonlinear spring and the dashpot were disconnected from the sink mass, resulting in a linear 2-DOF system which may be regarded as a baseline. Note that the response of the primary mass is essentially that of a viscously damped SDOF oscillator, whereas the sink mass response exhibits contributions from both the system's modes. Physically, the first mode of the combined system is strongly localized in the sink degree of freedom owing to the weak coupling provided by the linear spring (the coupling stiffness ε is approximately 1/8 the primary stiffness k).

The response of the nonlinear system formed when the cubic spring is connected between the sink mass and ground depends strongly upon the level of forcing. The accelerations of both masses following weak transient excitation (a light hammer blow) to the primary mass is shown in Fig. 9. The response of the primary mass is not substantially different from that of the linear system, and even the sink mass, which is directly attached to the nonlinear spring, behaves much as it did before.

When the transient force acting on the primary mass is larger, the nonlinearity of the cubic spring begins to influence the qualitative nature of the responses of both masses, as may be seen in Fig. 10. Most remarkable here is the envelope of the acceleration response of the primary mass. What was an exponential decay due to energy dissipation has become a much more rapid, even abrupt, decrease in vibratory amplitude due to energy pumping. A simultaneous large increase in the sink motion is easily observed in the lower trace.

By connecting the dashpot to the secondary mass, we provide a means to quickly dissipate the energy pumped to the sink. This effect is demonstrated by the response shown in Fig. 11. The excitation was a strong transient force acting on the primary mass. This figure shows clear evidence of energy pumping, with the majority of the energy initially imparted to the primary mass being transferred rapidly to the sink, where it was dissipated. This behavior was found to be repeatable.

5 CONCLUSION

A 2-DOF, nonlinear test rig has been designed and fabricated to demonstrate the phenomenon of energy pumping. This paper describes some of its unusual features, including the cubic-hardening, essentially nonlinear spring incorporated in the energy sink, and summarizes the results of measurements made of the apparatus itself. Energy pumping has been demonstrated in the laboratory using a specially designed 2-DOF test structure. The construction of this rig has been reviewed and data taken using it have been presented and discussed. The experimental occurrence of energy pumping was found to be robust.

6 REFERENCES

Gendelman, O. (2001), "Transition of Energy to a Nonlinear Localized Mode in a Highly Asymmetric System of Two Oscillators." *Nonlinear Dynamics*, **25**(1–3), 237–253.

Gendelman, O., Manevitch, L. I., Vakakis, A. F. and M'Closkey, R. (2001), "Energy Pumping in Nonlinear Mechanical Oscillators: Part I—Dynamics of the Underlying Hamiltonian Systems," ASME *Journal of Applied Mechanics*, **68**, 34–41.

Rivin, E. I. (1999), *Stiffness and Damping in Mechanical Design*, Marcel Dekker, Inc., New York.

Stoker, J. J. (1950), *Nonlinear Vibrations in Mechanical and Electrical Systems*, John Wiley & Sons, New York.

Vakakis, A. F. and Gendelman, O. (2001), "Energy Pumping in Nonlinear Mechanical Oscillators: Part II—Resonance Capture," ASME *Journal of Applied Mechanics*, **68**, 42–48.

Vakakis, A. F., Manevitch, L. I., Mikhlin, Y. V., Pilipchuk, V. N., and Zevin, A. A. (1996), *Normal Modes and Localization in Nonlinear Systems*, John Wiley & Sons, New York.

ACKNOWLEDGMENTS

This work was supported by Air Force Office of Scientific Research Grant Number 00–AF–B/V–0813 (Dr. Dean Mook, Grant Monitor) and Office of Naval Research Grant Number N00014–00–1–0187 (Dr. Luise Couchman, Program Manager). Mr. C. J. Hartwigsen and Mr. D. L. Gipson assisted in obtaining the experimental data reported herein.

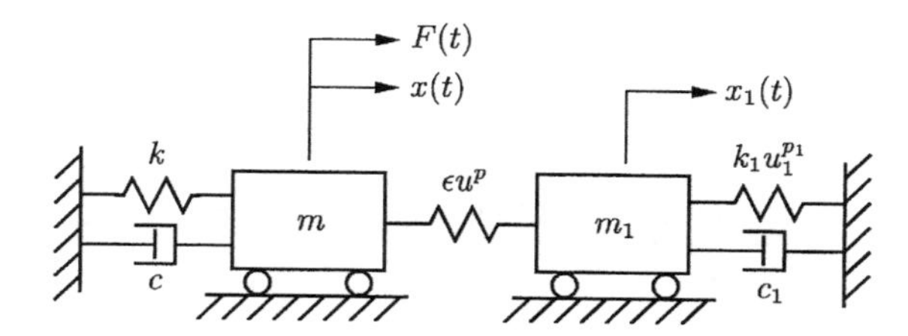

Figure 1: Two-degree-of-freedom system: SDOF primary structure plus NES.

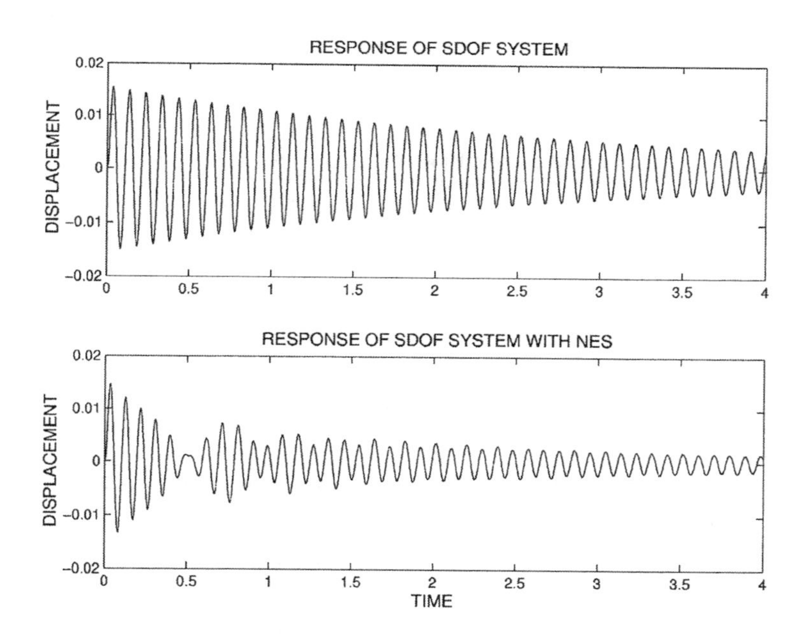

Figure 2: Simulated displacement response of primary system with and without energy sink.

Primary System	
m_1	1 kg
c	2.5 N/m/s
k	4000 N/m
Coupling Spring	
ε	400 N/m
Nonlinear Energy Sink	
m_1	0.5 kg
c_1	1.25 N/m/s
k_1	20×10^7 N/m^3
p_1	3

Table 1: Parameter values resulting in energy pumping in the simulated 2-DOF system.

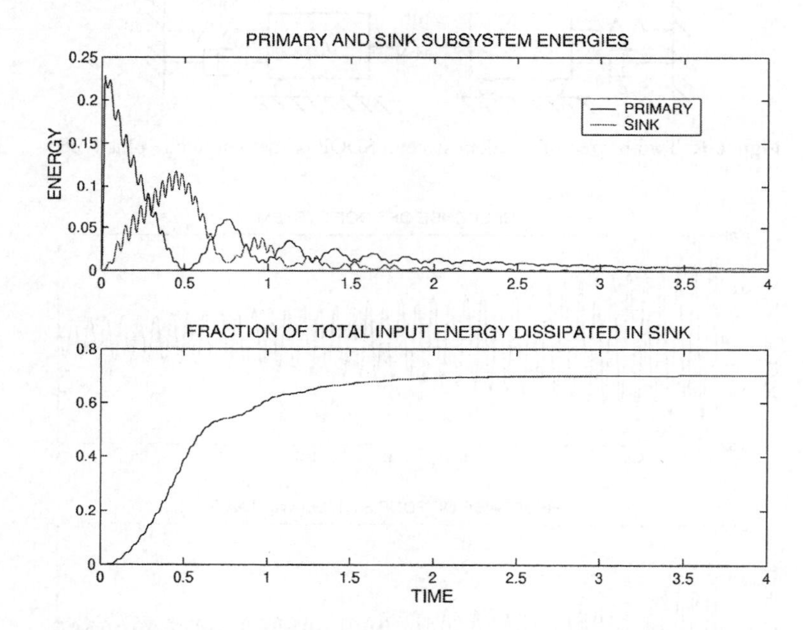

Figure 3: Energy distribution between primary and sink subsystems and dissipation in the damper.

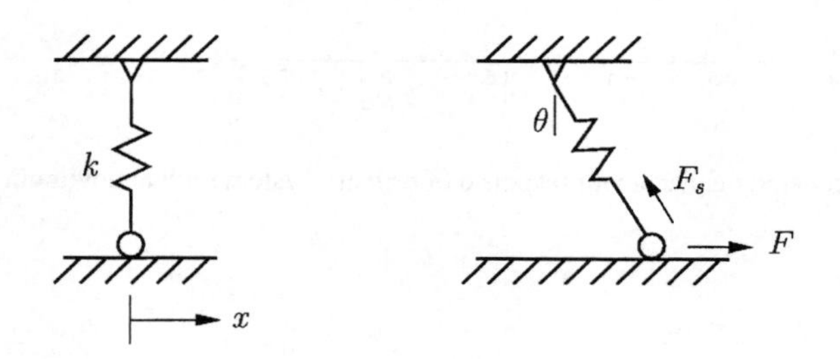

Figure 4: Geometry of spring deformation.

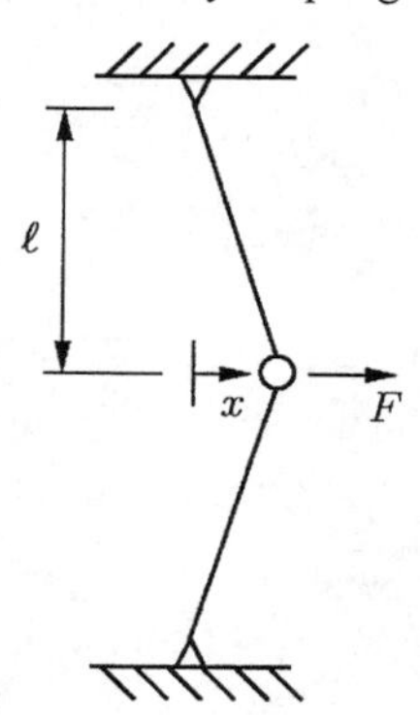

Figure 5: Practical essentially nonlinear spring design.

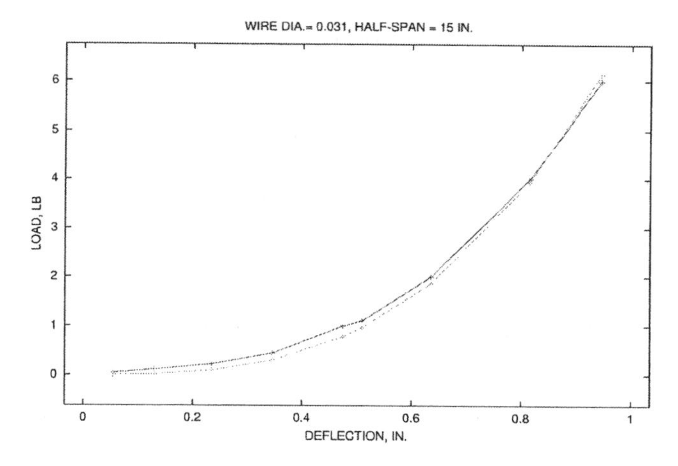

Figure 6: Force-displacement characteristic of the nonlinear spring.

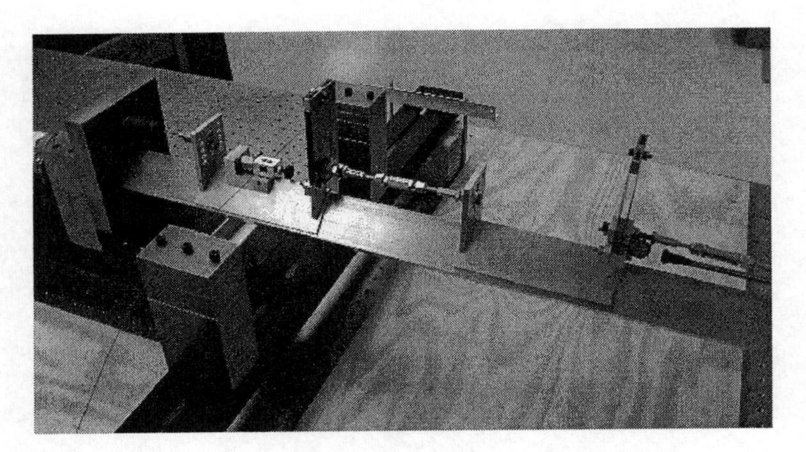

Figure 7: Two-degree-of-freedom energy pumping test rig.

Primary System	
m_1	0.510 kg
k	1060 N/m
Coupling Spring	
ε	130 N/m
Nonlinear Energy Sink	
m_1	0.234 kg
c_1	0.80 N/m/s
k_1	4×10^6 N/m^3

Table 2: Measured and identified values of the test apparatus.

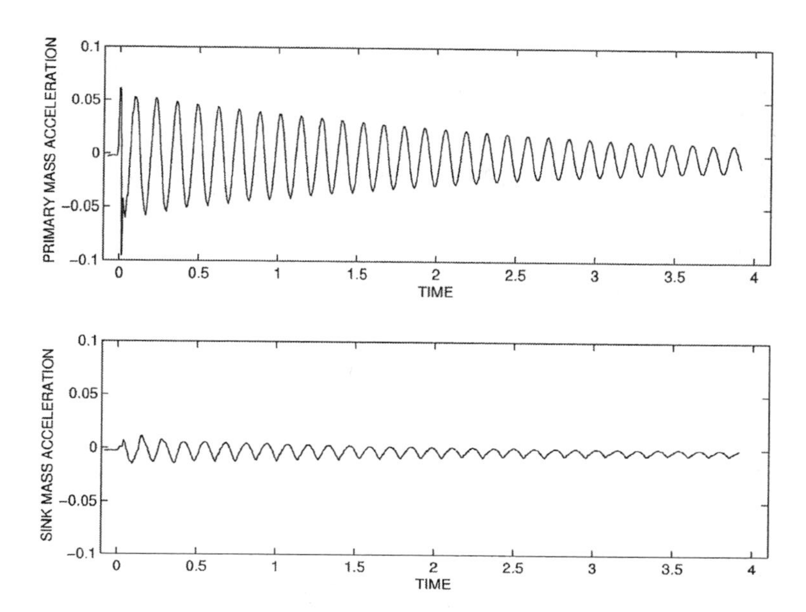

Figure 8: Response of the 2-DOF linear system (the nonlinear spring and the dashpot were disconnected during this test). Excitation was a hammer blow to the primary mass. (Experimental data.)

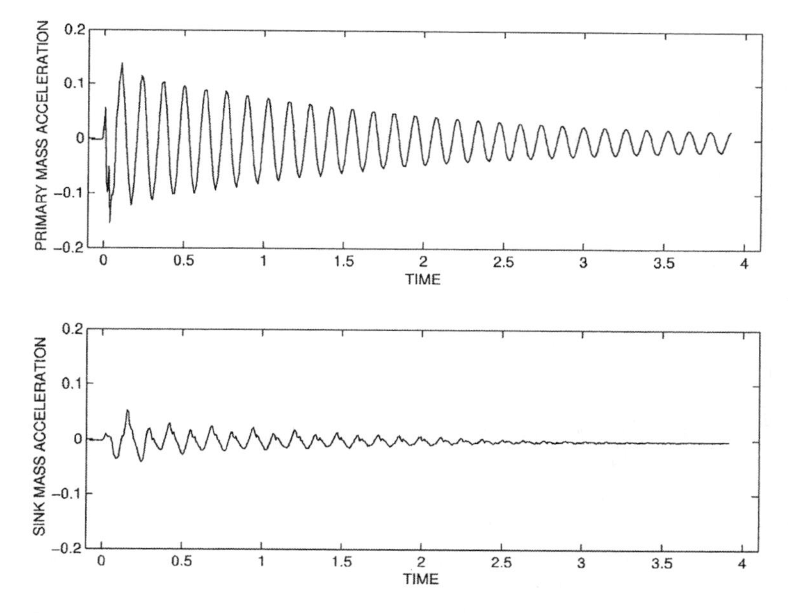

Figure 9: Response of the nonlinear system to weak transient forcing of the primary mass (dashpot disconnected). (Experimental data.)

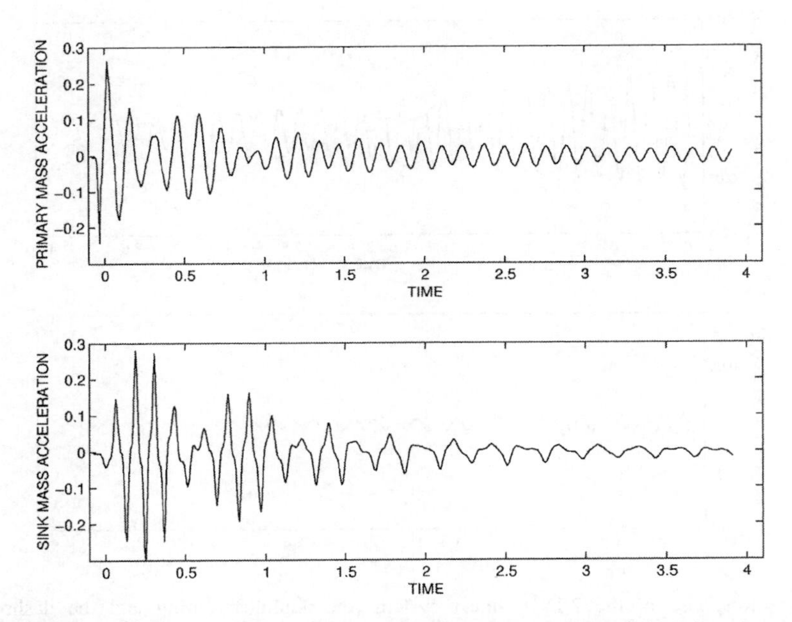

Figure 10: Response of the nonlinear system to strong transient forcing of the primary mass (dashpot disconnected). (Experimental data.)

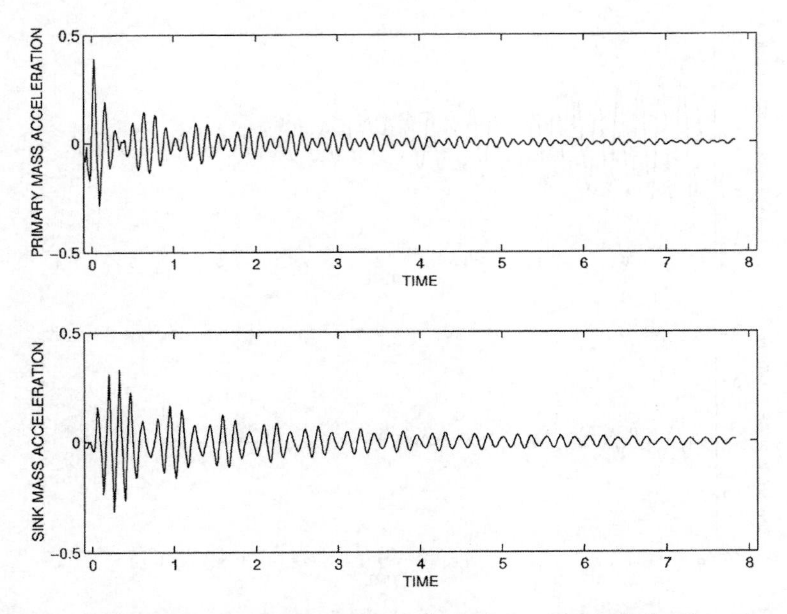

Figure 11: Energy pumping: Response of the nonlinear system (with dashpot) to strong transient forcing of the primary mass. (Experimental data.)

Signal Analysis and Artificial Intelligence in Structural Monitoring and Diagnosis

Alessandro De Stefano and Rosario Ceravolo

Dep. of Structural Engineering and Geotechnics, Politecnico di Torino, Italy

ABSTRACT

This paper reports on some new procedures to assess existing constructions and to predict damage scenarios at large territorial scale. In this connection, dynamic testing is an important tool, especially if it is oriented to global dynamics and experimental modal analysis. Classical structural identification procedures, which use transfer functions, require the knowledge or a reliable measure of the forcing actions, whilst some newly developed techniques, in particular time-frequency domain ones, exploit the natural excitation any structure is subjected to, and therefore do no require any special instrumentation. Structural identification techniques can be used as direct tools for modal parameter estimates, but it is usually more meaningful to associate them with F.E. models in automatic model updating procedures. Alternatively, the diagnostic problem can be reduced to the evaluation of functional symptoms reflecting the presence and the nature of the defect.

1. INTRODUCTION

The protection of a urban area or a life-line system against environmental offenses can generally be seen as a problem of optimal allocation of limited resources. By increasing the level of knowledge, it is possible to solve that problem more effectively. A robust and fast procedure to assess the vulnerability of existing construction and to predict damage scenarios at large territorial scale is useful to establish correct priority lists, so as periodic or continuous monitoring programs help managing important individual constructions.

Even non periodical experimental checks, if oriented towards diagnostics purposes, can supply a significant support to decision making for maintenance, retrofitting and control projects. The modern research has to face a fundamental challenge: designing procedures that are able to extract huge and reliable information with reduced costs and time demanding, both in vulnerability prediction and in experimental activities.

Dynamic testing is an important tool, especially if it is oriented to global dynamics and modal identification. A constant effort has been devoted, in recent years, to output-only characterization techniques. Time domain and frequency domain techniques usually require the input to be ergodic, or even a band-limited white noise. Such conditions seldom hold in real practice, hence the application of those techniques needs to be somehow forced to improper conditions.

Only time-frequency or evolutionary techniques can face effectively the case, so widely given, of non-stationary input. Wavelets can represent a direct and computationally efficient approach, due to their ability of being structured in sets of orthogonal filters and their simple implementation, but higher resolution, both in time and frequency, may be achieved with Cohen class t-f transforms, especially with the Choi-Williams kernels.

2. TIME-FREQUENCY IDENTIFICATION METHODS

2.1. Bi-linear estimators of modal parameters

Let us assume that the structure, subjected to unknown excitation, is instrumented with simultaneous acquisition channels according to some of the n degrees of freedom. In the time-frequency representation of the response signals, the energy appears to be concentrated around the modal frequencies and modulated according to the evolution of the time-frequency transform of the modulating waveform. Due to the fact that in the (t,f) plane the shape of the modulating waveform is maintained, it can be demonstrated that the amplitude ratio and phase difference between two measured signals $s_i(t)$ and $s_j(t)$ can be determined directly from their bilinear (Choi-Williams) time-frequency auto and cross representations:

$$D_{s_i}(t,f), D_{s_j}(t,f), D_{s_i s_j}(t,f)$$

in the following manner (Bonato et al 2000a):

$$AR_{i,j}(t,f) = \sqrt{\frac{D_{s_i}(t,f)}{D_{s_j}(t,f)}} \tag{1}$$

or, better:

$$PH_{i,j}(t,f) = \frac{D_{s_i s_k}(t,f)}{D_{s_j s_k}(t,f)} \tag{2}$$

where $AR_{i,j}(t,f)$ the time-frequency estimator for the modal amplitude, $PH_{i,j}(t,f)$ represents the time-frequency estimator for the modal phase difference and s_k is a pure sinusoidal signal at the previously detected modal frequency f.

2.1. Extracting Modal Parameters in t-f domain

2.1.1. Frequencies and Shapes

The phase difference estimator $PH_{i,j}(t,f)$ is a variable defined in the range 0-π. Given f, its SD_t

(Standard Deviation along the t axis) is generally not far from the mean value $\pi/2$.
It is not true if f is a modal frequency; in that case, the phase difference between simultaneous records tends to a constant value, scattering is suddenly decreasing and $SD_t[PH_{i,j}(t,f)]$ falls down to zero. On the $SD_t[PH_{i,j}(t,f)]$ versus frequency plot, modal frequencies are revealed by sharp downward peaks.

If the unknown input is a random non-stationary process, then different modes can be variously excited in different time windows. It is a powerful property of Time-Frequency distributions, useful to easier separate closely coupled modes.

The described procedure uses local cross spectra, instead of auto-spectra, as most of the T-F related approaches do. It allows to clean out the uncorrelated random noise, present in each individual signal.

Once detected each f_p modal frequency the corresponding modal shape is supplied by the square matrix $\boldsymbol{AR}_p$, whose general term id the average along the time axis of the amplitude ratio estimator $AR_{i,j}(t,f_p)$.

2.1.2. Damping

At first sight, the extension of the results discussed above to the evaluation of damping may appear direct. Actually, in this case new problems arise:
- Damping affects the modulation of the signal rather than its instantaneous frequency;
- In practical applications damping evaluation is critical, and any available data about the system's behaviour should not be neglected by the identification procedure.

Based on the foregoing considerations, the proposed method includes the following stages (Ceravolo et al 2001):
- Time-frequency transformation of the response signals and regularization. The latter point is achieved by imposing, at any time t, the linear model formulation in the frequency domain, or, in other words, by searching the Frequency Response Functions that best fit the instantaneous response spectra. For each modal frequency, diagrams showing the instantaneous estimate of damping are plotted.
- The final value of the damping is obtained by averaging over time the instantaneous damping diagrams. Results coming from different channels should be averaged, or combined by resorting to multi-criteria techniques.

The idea of assigning an instantaneous nature to damping is not questionable from a physical point of view, but it produces many counterintuitive consequences. In fact, the instantaneous response spectrum is conditioned by the signal's behaviour near the time t, though, from a strictly energetic point of view, a change in the damping factor would affect only the following part of the signal. On the other hand, if the input is unknown, the local information about energy is missing. We can conclude that the instantaneous damping defined here is only indirectly related to the physical concept of dissipation, but rather it expresses some local modulation characteristics of the response signal. In the proposed method, these characteristics are compared instant by instant with those exhibited by similar systems in stationary conditions, in order to extract the most likely local damping coefficient. Hence, for a single degree of freedom (SDOF) system, the value of the parameter ζ that minimizes the following functional is the estimate of the instantaneous damping (Ceravolo et al 2001):

$$\zeta(t)=\min_{\zeta}\left[E(t)=\int_f\left[D_{s_i}(t,f)-\left|H(f,\zeta)\right|^2\right]df\right] \tag{3}$$

In this minimization, there is an explicit assumption that the instantaneous spectrum $D(t,f)$ approaches a scaled version of the Frequency Response Function, $H(f)$, whose maximum values at the modal frequencies are determined directly from the peaks of the time-frequency distribution. A similar relationship could be written for velocity or acceleration signals, and eq. 3 may easily be generalized to multiple degrees of freedom (MDOF) systems, which require more damping parameters to be optimized.

One of the advantages of using eq. 3 is that it forces the regularized instantaneous spectrum to satisfy the time marginal condition:

$$\left|s_i(t)\right|^2=\int_f D_{s_i}(f,t)df=\int_f\left|H(f)\right|^2_{\zeta=\zeta(t)}df \tag{4}$$

This means that the regularisation operation does not alter the instantaneous energy. Conversely, the frequency marginal, i.e. the spectral energy, which retains great significance in stationary signals, has been sacrificed.

2.2. Example of application to a real structure

The first case study reported is the SS. Annunziata church bell-tower in Roccaverano (Bonato et al 2000b). A vibration test campaign was carried out on a XVI century church bell-tower rising in Roccaverano (Asti-Italy) whose style is inspired to the school of the famous renaissance architect Bramante (Figure 1). In the past the church was exposed to a strong earthquake which caused serious damage both on the facade and on the bells-tower and subsequently some interventions of restoration were made. The tower has been subjected to an extensive experimental investigation both under ambient vibrations and actions induced by the bells.

Vibration measurements were performed on the bell tower only, by placing the accelerometers on the landings, arranged in the horizontal direction. Each set-up is made up of the signals relating to four acquisition channels, of which two were fixed as reference channels and two were moved to the levels of the different landings. The measurements were made separately in the E-W and N-S directions, but one of the tests was conducted with two accelerometers arranged in the orthogonal direction, according to the two main axes of the bell tower, in a central position. This made it possible to correlate the modal shapes observed in the two main directions and to build up space modal shapes, of special importance in connection with mainly torsional modes.

Different types of excitation were used, and namely, the one generated by bell tolling in two different directions, the one produced by pulses applied to the bells and finally the one arising from environmental noise. An aspect to be noted is the absolute absence on Fourier spectra of important components in the 2.5-10 Hz range, namely a frequency range within which the second flexural mode and the first two mainly torsional modes are typically located in structures of this type. Details of this application, summarized in Figures 1-3 and Table 1, may be found in Bonato et al (2000b).

Figure 1. The bell-tower

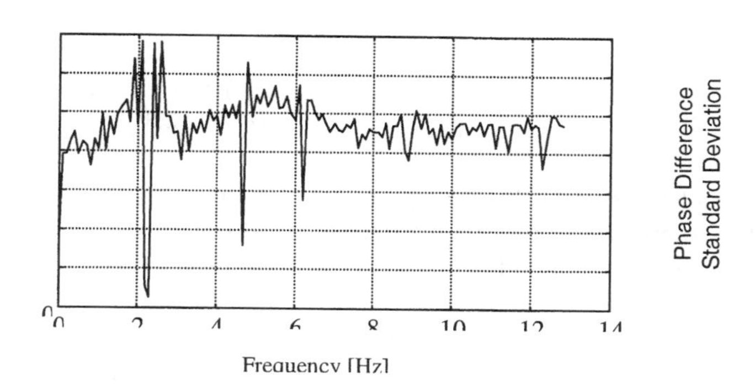

Figure 2. The $SD_t\,[PH_{i,j}(t,f)]$ function

Table 1. Detected modal frequencies

Mode	Identified frequencies (Hz)
1	1.66
2	2.26
3	4.67
4	6.18
5	6.40
6	8.90

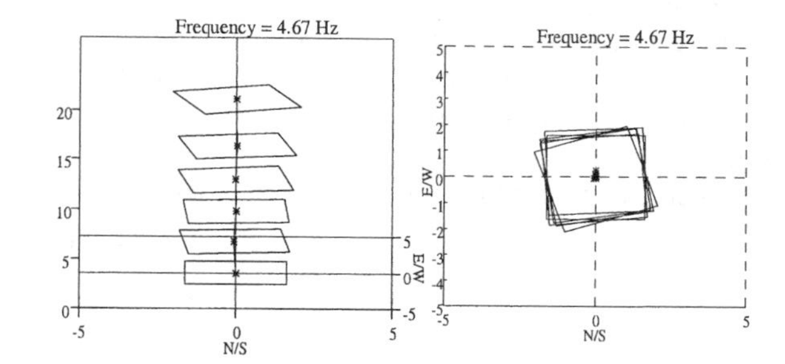

Figure 3. A sample mode identified in the range 1-5 Hz (1st torsional)

3. DIAGNOSTIC APPLICATIONS

The use of ambient, output only, data makes it mandatory to have multiple information sources in order to extract direct reliable diagnostic warnings. A.I., and specially Neural Networks, can supply a fast, robust tool to achieve that goal.

3.1. Detection of non-linearities

Structural diagnosis can be performed according to the two different strategies (Ceravolo et al 1996):

- knowledge of the undamaged state and periodical comparisons with current conditions.
- identification of the "symptomatic" characteristics of structural response.

The latter strategy is indispensable in the diagnosis of existing structures. The symptoms may reflect irregularities in terms of instantaneous frequencies, damping and modal forms. Above all, the non-linearity of the structural response often proves very significant as a telling symptom. This paper reports on a technique for the identification and classification of structural non-linearities which uses the Hilbert transform in the domain of time (De Stefano et al 1997).

The Hilbert transform in the frequency domain is obtained from the convolution integral:

$$H(\omega_c) = -\frac{1}{i\pi} PV \int_{-\infty}^{+\infty} \frac{G(\omega)d\omega}{(\omega - \omega_c)} \tag{5}$$

being PV the " Cauchy principal value of the integral".

This transform can be used to detect non-linearities in the frequency response function G(ω) according to the following criterion (Tomlinson 1987):

$$H(\omega) = G(\omega) \Rightarrow \text{linear system}$$
$$H(\omega) \neq G(\omega) \Rightarrow \text{nonlinear system}$$

This approach proves particularly effective in the interpretation of dynamic tests with sinusoidal type excitation. The main difficulty in the application of this technique is that the frequency response function (FRF) and hence the excitation are not always known.

The Hilbert transform in the time domain makes it possible to identify the modal parameters and the non-linear characteristics of a dynamic system even when the excitation is not known. Of special significance, in this field, are the works by Feldman (1994) and Spina et al (1996).

The Hilbert tranform of the signal, expressed by:

$$H[y(t)] = \tilde{y}(t) = \frac{1}{\pi} \int_{-\infty}^{+\infty} \frac{y(\tau)}{t - \tau} d\tau \tag{6}$$

is added to the signal itself to form the analytical signal, as follows:

$$Y(t) = y(t) + j\tilde{y}(t)$$

The cyclic paths of this representation, projected onto the complex plane, are referred to as "orbits" (Figure 4). This approach makes it possible to detect in a clear and direct manner the linear and non-linear characteristics of structural response. In particular, the representation of the structural response in the space delimited by the axis of time and the complex plane of the analytical signal supplies a direct reading of the influence of non-linear characteristics on the response of a vibrating system and the evolution of such influence as a function of time.

From Figure 5 it can be seen that each type of non-linearity taken into account characterizes in a different manner the shapes of the orbits produced by the analytical signal over time.

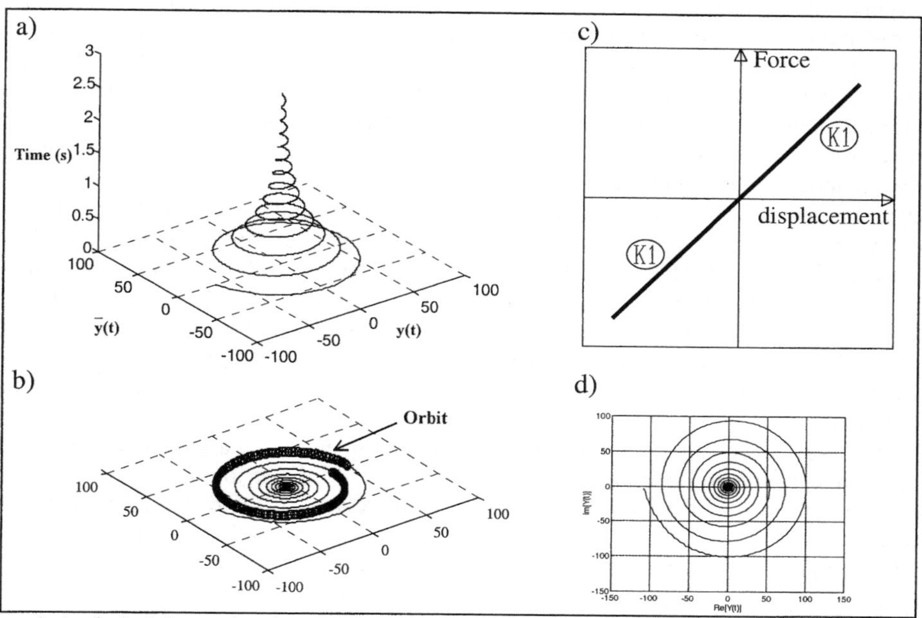

Figure 4. Analytical linear signal representation: a) 3D image (complex plane vs time); b-d) projections of analytical signal trace on complex plane (one "orbit" bold); c) force-displacement plot (De Stefano et al 1997)

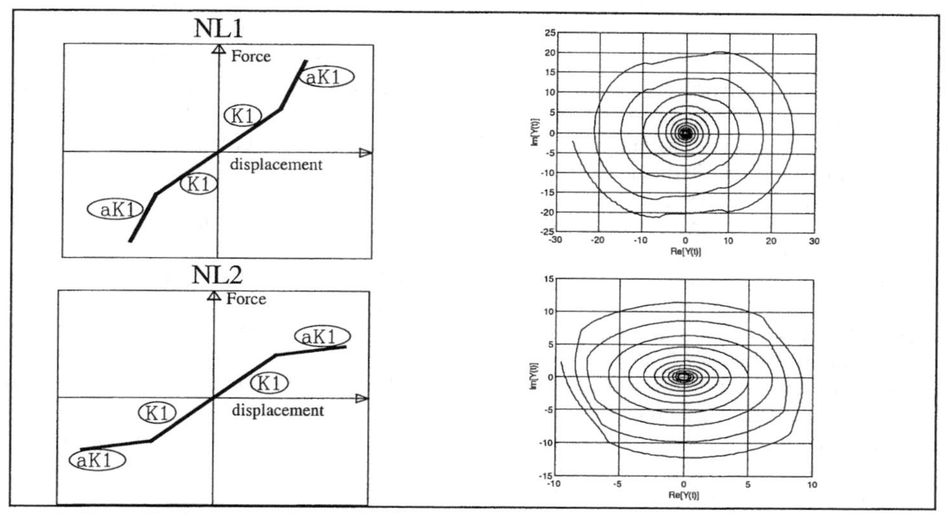

Figure 5. Force-displacement plots and related analytical signal traces on complex plane in different non-linear behaviour (De Stefano et al 1997).

4. CONCLUSIONS

Time-frequency techniques for structural dynamic output-only identification are effective in supplying a direct modal frequency and shape estimation and they seem to be promisingly effective in case of damping estimation. Moreover, they do not need any assumption of input stationarity. This allows to process signals with short data sets, whilst time-domain and frequency domain methods generally demand long records. All direct structural identification techniques have the problem of incompleteness and can be used as direct tools for modal parameter estimates, but it is usually more meaningful to associate them with F.E. models in automatic model updating procedures.

Dynamic testing can be useful also in symptom based damage detection techniques. The problem can be reduced to the evaluation of functional symptoms reflecting the presence and the nature of the defect. Therefore, it is necessary to identify a suitable set of input conditions and output parameters and to determine the number of experiments to be performed in order to achieve an unequivocal identification of the damage. It is common practice to resort to a great number of input conditions and to program series of experiments as extensive as possible. These methods do not offer predictive tools, but they appear very promising in terms of model recognition .

In other words, with simple model recognition techniques signals are treated as patterns that are attributed to specific damage classes. On the basis of this attribution, there is a direct or indirect subdivision of the signal space into regions related to the different classes. Pattern recognition techniques - often combined with a joint representation in the time-frequency domain - can be used most effectively to realise this approach. Neural networks can be a very helpful tool in practical approaches to pattern recognition.

REFERENCES

Bonato P, Ceravolo R, De Stefano A, Molinari F. (2000) Cross-Time-Frequency Techniques for the Identification of Masonry Buildings *Mechanical Systems and Signal Processing*, 14: 91-109.

Bonato, P., Ceravolo, R., De Stefano, A., Molinari, F. (2000) Use of cross Time-Frequency Estimators for the Structural Identification in Non-Stationary Conditions and under Unknown Excitation, *J. Sound and Vibration*. 237: 775-791.

Ceravolo, R., De Stefano, A., Molinari, F. (2001) Developments and Comparisons on the Definition of an Instantaneous Damping Estimator for Structures under Natural Excitation *Key Engineering Materials*. 204-205: 231-240.

Ceravolo, R., De Stefano, A., Sabia, D (1995) Hierarchical Use of Neural Techniques in Structural Damage Recognition *Smart Materials and Structures*. 4: 270-270.

De Stefano, A., Sabia, D., Sabia, L. (1997) The Use of Hilbert Transform and Neural Intelligence in Structural Non-Linearity Detection, *J. Structural Control*. 1: 89-105.

Feldman, M. (1994) Non-linear System Vibration Analysis Using Hilbert Transform", *Mechanical Systems and Signal Processing*. 8: 119:127.

Spina, D., Valente, C., Tomlinson, G.R. (1996) A New Procedure for Detecting Nonlinearity from Transient Data Using the Gabor Transform. *Nonlinear Dynamics*. 11: 235-254.

Tomlinson, G.R. (1987) Developments in the Use of the Hilbert Transform for Detecting and Quantifying Non-linearity Associated with Frequency Response Functions, *Mechanical Systems and Signal Processing*. 1: 151-171.

Use of Nonparametric Approaches for Structural Health Monitoring of Dampers

Raymond W. Wolfe[1], Sami F. Masri[2], and John Caffrey[2]

[1]*Supervising Bridge Engineer, California Department of Transportation, Sacramento, CA*

[2]*Department of Civil Engineering, University of Southern California, Los Angeles, CA*

ABSTRACT:

Viscous dampers are finding increasing use in the retrofit or design of new bridges subjected to strong dynamic loads such as earthquakes. Due to the long service life and harsh environments that such dampers may be subjected to, it is important to have a reliable and efficient procedure for condition assessment of these dampers based on the analysis of their vibration signature.

This work presents an overview of a promising approach based on nonparametric identification techniques to detect slight changes in the characteristics of nonlinear viscous dampers of the type commonly encountered in structural control applications involving large civil infrastructure systems such as bridges. By characterizing the restoring force surface of the damper in a nonparameteric form, and subsequently analyzing the higher-order statistics of the coefficients defining such surfaces, it is found that the associated probability density function of the identified coefficients furnishes a sensitive indicator of the underlying damper parameters. Both simulation results as well as experimental measurements from a prototype nonlinear viscous damper are presented to illustrate the approach and discuss its range of validity.

1. INTRODUCTION

The work reported in this paper is drawn from analytical and experimental studies designed to evaluate a promising approach for the detection of subtle variations in structural system parameters utilizing higher-order statistics. The identification algorithm used for data processing is based on a Chebyshev polynomial representation of the simulated or measured system response. This approach yields a set of nonparametric coefficients, representative of the system physical properties. A brief overview of simulation studies is presented in Section 2. Experimental studies, including temperature effects are presented in Section 3.

2. SIMULATIONS

2.1. Model Development

Simulations were performed using linear and then nonlinear single-degree-of-freedom oscillators. The results of a Duffing oscillator with a unit mass and period, and a natural frequency of 2π, and nonlinear term ε of 11.23 are presented herein. The restoring force for a Duffing oscillator can be written as

$$f(x,\dot{x}) = m[2\zeta\omega\dot{x} + \omega^2(x + \varepsilon x^3)] \tag{1}$$

The resulting system stiffness from the above system description was 39.48, and the damping coefficient was 1.26. The simulation data was corrupted with stationary, zero-mean noise having a standard deviation of 0.10 to account for noise sources such as instrumentation susceptibility, cabling interference, acquisition hardware, etc.

2.2. System Identification

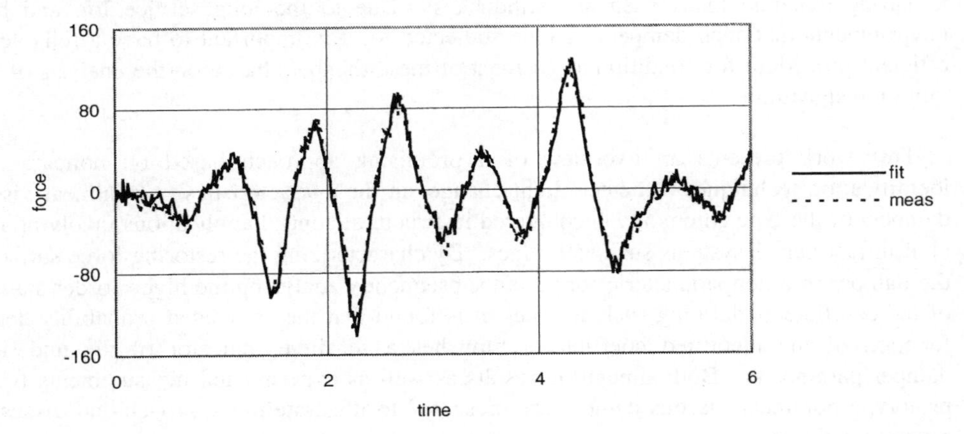

Figure 2.1 Comparison of measured and identified restoring force

The restoring force identification algorithm yields normalized Chebyshev coefficients. A 3rd order Chebyshev approximation was assumed for the simulation of this nonlinear system. A comparison of the measured and identified system restoring is illustrated in Figure 2.1. Clearly, the identified force closely matches the measured value, illustrating the identification algorithm's robustness in detecting nonlinear systems under noise pollution.

A three dimensional representation of the identified restoring force surface is presented in Figure 2.2. The system nonlinearity is evident in this figure as a linear system would present a flat plane sloping in the displacement-force axes, with the slope describing the system stiffness. The surface illustrated in Figure 2.2 reveals peaks and valleys in several orientations, highlighting the nonlinear Duffing oscillator response.

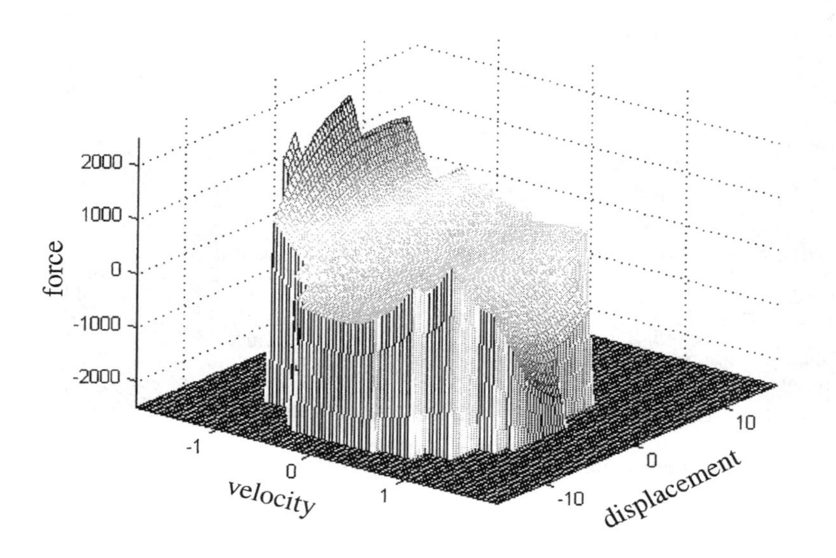

Figure 2.2 Identified restoring force surface from simulated model, random excitation

3. EXPERIMENTAL STUDIES

3.1. Test Description

Experimental studies were performed with a scaled prototype nonlinear viscous damper. The damper was a scaled version of the seismic dampers currently in service or being installed on a number of large suspension bridge structures in California. The tested damper had a 10 kip force rating at 12-inches displacement and a velocity rating of 70 ips. To perform the testing, a test facility was constructed with an 11-kip MTS Systems actuator. The 90 gpm servo valve on the actuator was sized to accommodate the high damper velocity rating, yielding excitation frequencies of 30 Hz for ½ inch displacement amplitudes. The damper/actuator connection was fabricated with linear bearings to only allow motion along the axis of the damper, thereby avoiding out-of-plane bending. A photograph of the test setup is provided in Figure 3.1.

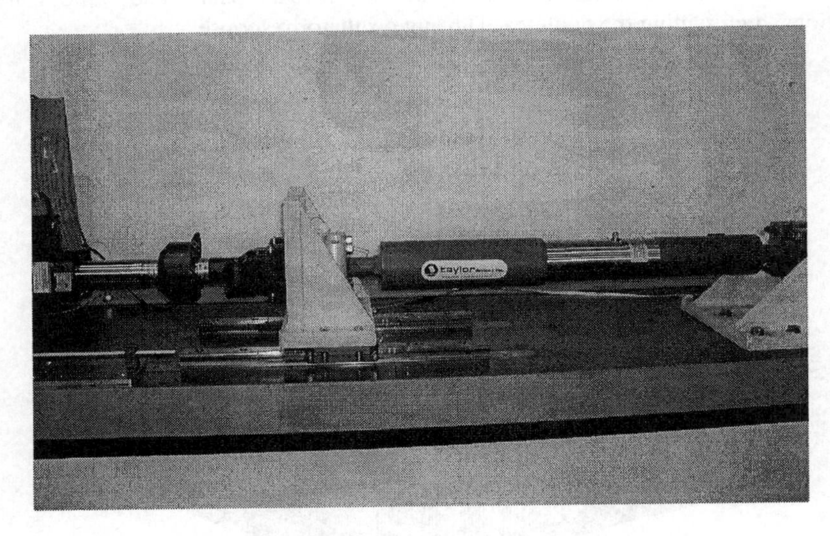

Figure 3.1 Photograph of experimental test setup

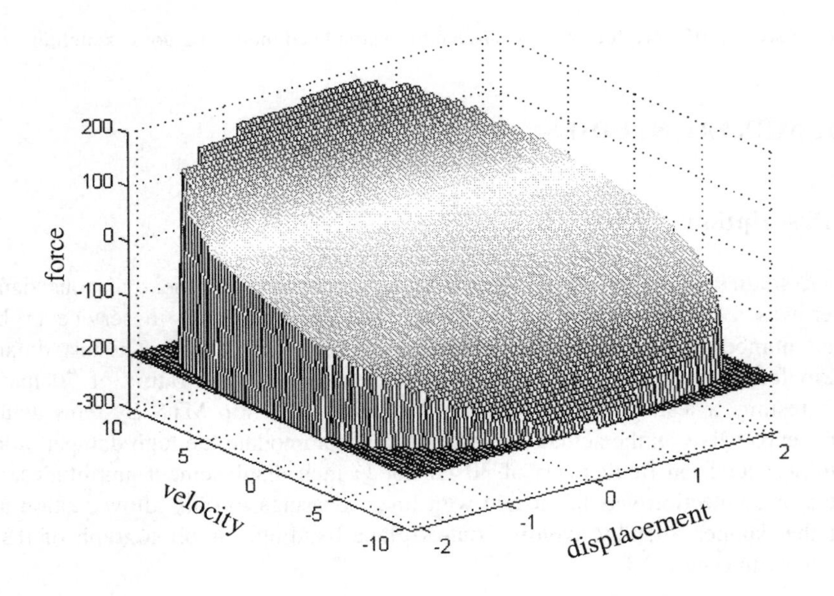

Figure 3.2 Estimated restoring force surface from measured data, sinusoidal excitation

3.2. Experimental System Identification

Direct measurements of displacement, velocity, acceleration, force and exterior damper temperatures were recorded during testing. The measured system response was processed through the restoring force identification algorithm to determine the test article stiffness and damping parameters. This method computes a mathematical representation of the restoring force surface in terms of Chebyshev polynomials. A relatively straightforward transformation yields a representative Power Series system response approximation (Wolfe, 2002). An example of the restoring force surface is depicted in Figure 3.2 for a system excited by a 4 inch-peak 1 Hz sinusoid with the damper initially set between 90-100 °F.

3.3. Temperature Effects

Temperature effects were studied by performing tests at three different temperature ranges: T1: 70 – 80°F, T2: 90-100°F and T3: 110-120°F. Variations in the measured system response resulting from temperature effects are presented in Figure 3.3. Figure 3.3 depicts the measured restoring force for each of the three cases. The identified Chebyshev polynomial coefficients for the reference case T1 is included in Table 3.1. These experiments were performed with a sinusoidal excitation having a 4 inch-peak amplitude at 0.5 Hz.

The measured damper force response illustrated in Figure 3.3 reveals that environmental forces such as temperature variations do affect the system response. The frequency remains fairly constant as expected (the plotted shift is due to the starting point of the test), but the amplitude is reduced by 5% at T2 and 12% at T3, compared to the measured response at T1.

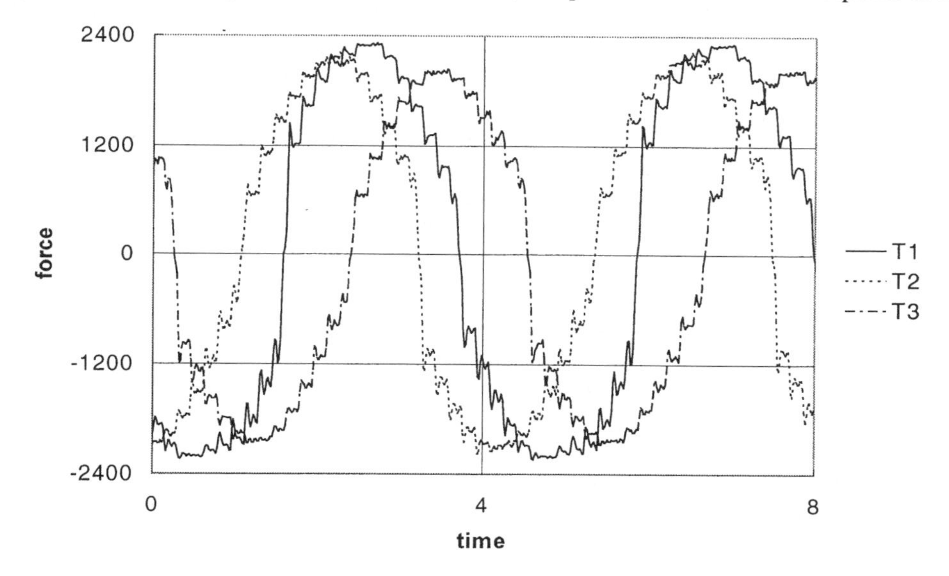

Figure 3.3 Measured restoring force comparison, sinusoidal excitation

The reported temperatures were measured external to the damper, so do not reveal the true internal fluid temperature. The test was designed in this manner to facilitate simple field application with existing in-service dampers of similar design. Further testing or finite element analysis to define the thermal gradient for different damper cross-sections would be necessary to apply these results to other damper designs.

Chebyshev coefficients (T1)				
i/j	0	1	2	3
0	-194.98	-2460.62	30.98	138.83
1	-11.45	-12.96	-0.43	22.81
2	-31.92	18.03	5.92	0.48
3	0.78	0.56	1.46	-2.74
Power series coefficients (T1)				
i/j	0	1	2	3
0	-171.09	-893.54	3.84	16.29
1	-4.09	-15.74	-0.53	1.73
2	-17.07	2.30	0.51	0.02
3	-0.31	1.15	0.13	-0.14

Table 3.1 Identified Chebyshev polynomial coefficients

CONCLUSIONS

Simulation and experimental results were presented in this paper to demonstrate the validity of the restoring force algorithm in detecting changes to system parameters subjected to environmental factors. Noise pollution effects were presented with the simulation work. The effects of temperature on a scaled viscous damper representing large-scale devices currently being employed on several long-span structures in California and elsewhere were discussed. The restoring force method has been shown capable of detecting subtle system variations from noise-polluted data sets. Additional simulation and experimental work supporting these conclusions is detailed in Wolfe (2002).

References

Masri, S.F., and Caughey, T.K., (1979), "A Nonparametric Identification Technique for Nonlinear Dynamic Problems", ASME *Journal of Applied Mechanics*, Vol. 46, June 1979, pp. 433-447.

Wolfe, R.W., Masri, S.F., and Caffrey, J., (2002), "Some Structural Health Monitoring Approaches for Nonlinear Hydraulic Dampers", *Journal of Structural Control*, February 2002.

Wolfe, R.W., (2002), "Analytical and experimental studies of structural health monitoring of nonlinear viscous dampers", Ph.D. Dissertation, University of Southern California, May 2002.

Seismic Response of Elevators in Buildings and Their Sensor Needs

M. P. Singh and Rildova

Department of Engineering Science and Mechanics, Virginia Tech, Blacksburg, VA 24061 USA

ABSTRACT

There is a large inventory of elevators in seismically active areas. The elevator systems, and especially their counterweights, are quite vulnerable to earthquake induced ground motions. The paper presents some results of a numerical study, which considers nonlinear flexibility characteristics of the guide rails, brackets, and roller guide assemblies to investigate the vulnerability of these systems in seismic events. The paper also discusses the data collection methods that are currently used during and after an earthquake. The need for a more comprehensive data collection system is emphasized and the use of new sensors is proposed.

INTRODUCTION

There are several components in an elevator that can get damaged during an earthquake, but the counterweight being the heaviest is the most vulnerable of all the components. To analytically ascertain the dynamic performance of an elevator system in an earthquake, it is essential to prepare an accurate model of the counterweight system recognizing its all important features. The counterweight is guided up and down the rails as a rigid body by roller guide assemblies. A roller guide assembly consists of three or more rollers that are kept in constant contact with the guide rails by pre-loaded springs. To improve riding quality and the conserve energy, the rollers are provided with elastomer tires. The past analytical studies have shown that the dynamics of a counterweight system is strongly affected by the flexibility of the roller guide components as well as the flexibilities of the guide rail and bracket systems. The dynamic behavior of a counterweight is also affected by the gaps and clearances between the guide rails and the frame and the guide rail and the restraining plate that are used to confine the roller guide assemblies near the rails. Closing of these gaps introduces nonlinearity in the systems, and it is important to include this in the dynamic analysis of these systems. The possible nonlinearity due to yielding of the rails during a strong earthquake is not of interest as by then the derailment of the counterweight and the damage of the system is inevitable.

The nonlinear force deformation characteristics of the composite spring representing the flexibility of the roller guide assemblies and the rail-bracket systems are shown in Figure 1a. The initial flatter portion of the diagram pertains to the case when the gap between the restraining plate and the rail is still open, and the steeper part of the diagram is realized when the gap is closed. Similar force deformation diagram, shown in Figure 1b, is realized when there is a contact between the rail at the bracket support and the counterweight frame.

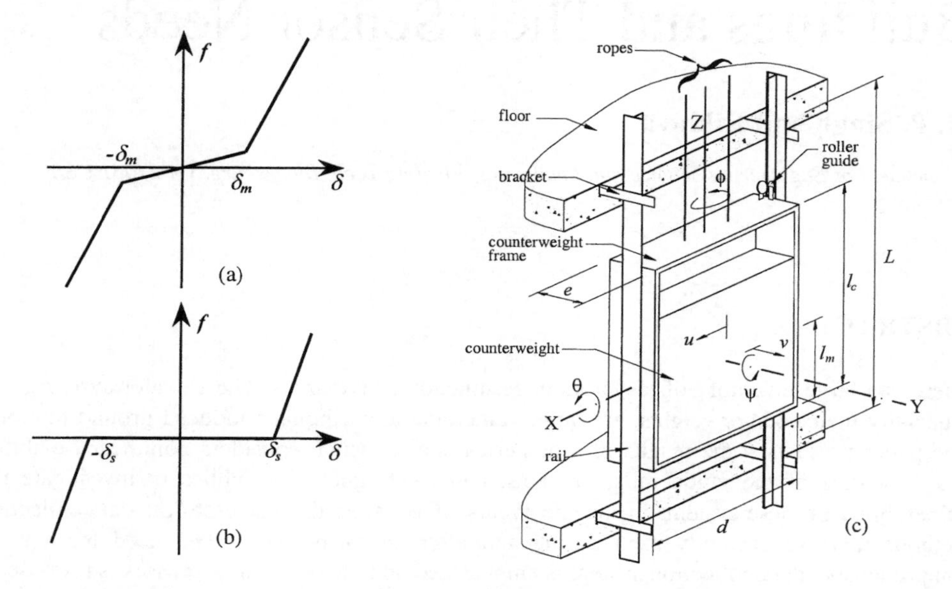

Figure 1. The bilinear restoring forces for (a) equivalent spring at each, and (b) contact between rail and counterweight frame; and (c) geometry of the 3-dof model of the rail-counterweight system

NUMERICAL ANALYSIS AND RESULTS

Figure 1c shows the schematics of an elevator counterweight with the degrees of freedom associated with its motion. For these degrees of freedom, the equations of motion for the out-of-plane and in-plane motions can be easily developed; details of these are given by Singh et al. (2002). Some seismic response results calculated for a counterweight of an elevator in a 10-story building are presented. The solution of the counterweight equations of motion required special attention because of the nonlinear characteristics of the system springs (Figures 1a and b). The fourth order Runge-Kutta scheme with adaptive time steps was found to be most efficient and accurate, and used here for the analysis of such a system.

Figure 2 shows the maximum stresses in the web of the rails as the counterweight traverses along the building height. Also shown are the forces in the brackets and the maximum floor acceleration in the X-and Y- directions. The input consists of the two components of the Northridge Earthquake, with larger component normalized to a maximum ground acceleration value of 0.1g. The brackets experience large impact forces when the gap between the restraining plane and the rail near a bracket support gets closed during the vibration of the system. The maximum stresses and forces need not occur near the top of the building even though the accelerations are the largest there. The stresses and forces, of course, increase with the intensity of ground motion. In Figure 3a, we show the stresses in the 18.5-lb rails for the recorded Northridge and El Centro ground motions when the counterweight travels down the top story. Figure 3b compares the stresses in the 18.5-lb and 30-lb rails for the recorded Northridge motion.

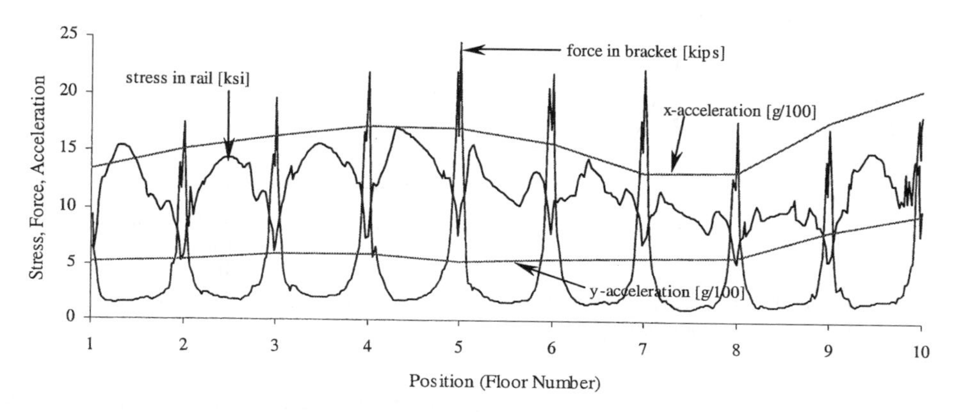

Figure 2. Maximum stress in the rail as a function of counterweight position along the building height, Northridge 0.1g, 18.5-lb. rail

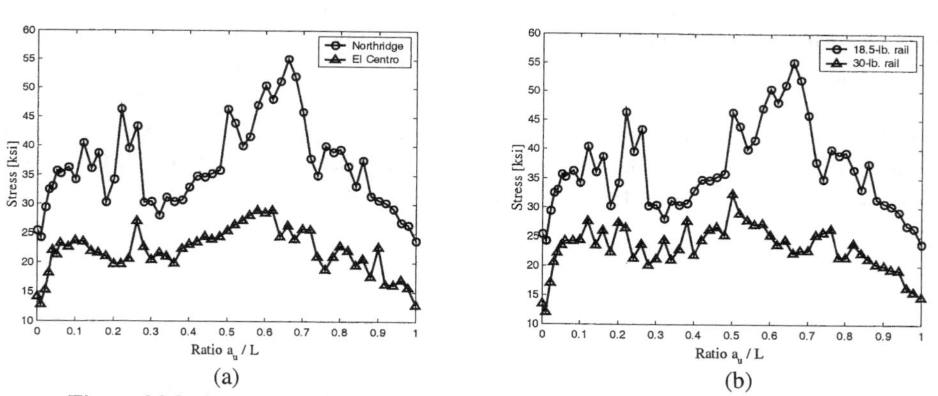

Figure 3 Maximum stress in the rail for (a) 18.5-lb. rail; (b) Northridge earthquake.

Other system characteristics that affect the response are the gap sizes. The code requires a clearance of no more than 3/16 inch at the restraining plates, and ½ inch between the rail and the frame. Figures 4a and 4b show the effect of changing these clearances on the maximum stresses in the rails. The stresses shown in these figures are the maximum of the stresses at seven critical locations in the rails. The stresses increase with the gap size until the gap becomes too large to cause any contact between the components. The stresses in Figure 4b for the El Centro motion are low primarily because of the intensity of the motion is smaller.

When the counterweight is in the two adjacent spans, a reduced clearance brings the frame in contact with the support more often. This reduces the bending in the rails, but increases the forces on the brackets.

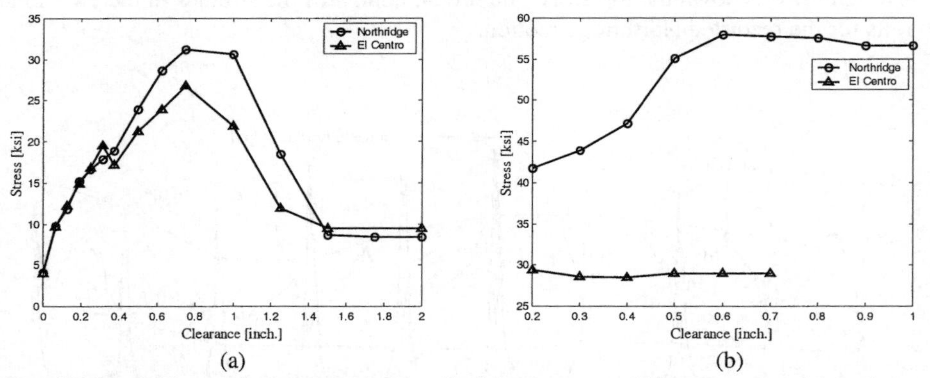

Figure 4 Effect on the maximum stress in the rail of the clearances (a) between restraining plate and rail, (b) between counterweight frame and rail

The code also recommends the use of two or more tie brackets between the rail bracket supports to reduce the in-plane bending effect. The numerical results obtained here confirm this effect of the tie bracket. With a tie bracket in the middle, the maximum stresses in the rail were reduced by 52% for the Northridge earthquake and by 44% for the El Centro earthquake. There were also some increases in the stresses (15% Northridge and 5% El Cenro) when the counterweight was near one of the supports, but this increase was not as significant as the decrease mentioned above. Figure 5a and 5b show this effect on the fragility of the rails calculated in the middle and near the end. This fragility estimation is based on a simulation study involving an ensemble of 50 synthetically generated ground motions. A dramatic reduction in the fragility in the middle but some degradation in the fragility near the bracket support due to installation of the tie bracket is noted.

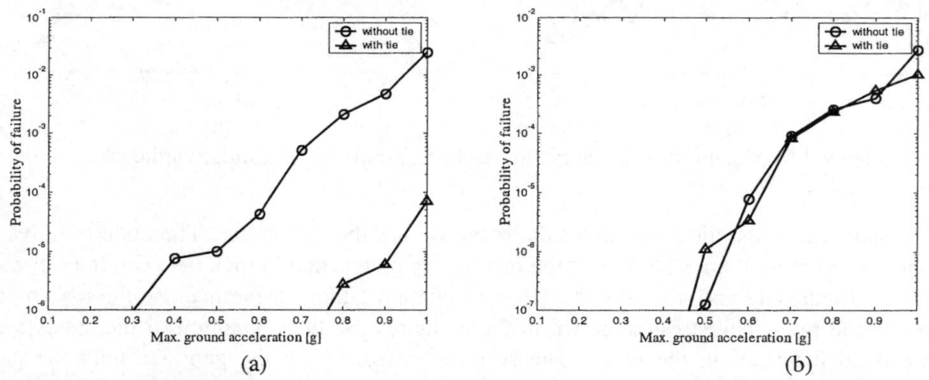

Figure 5 Effect of tie-bracket on the fragilities (a) when counterweight is in the middle of the rail, (b) when the counterweight is 0.25L from the top support

SENSOR NEED FOR ELEVATORS

We note that elevators are complex mechanical and electrical components with complex dynamic characteristics that affect their response in a complex manner in a seismic environment. There has not been any past earthquake events of engineering interest in which elevator damages have not been reported. The information about the damage began to be collected and reported only after the 1964 Alaska earthquake and in a more systematic manner after the 1971 San Fernando earthquakes. Most of these reports usually provide a numerical tally of damages, with some qualitative descriptions. Although these statistical summaries have been instrumental in improving the elevator codes, they lack relevant engineering details that are needed to establish the real cause of the failures. Since the post-earthquake reconnaissance teams usually do not have access to the damaged elevator components in the cramped hoistway areas, the repair service personnel usually do the task of collecting failure information whenever the repairs are made. In the understandable rush to restore service, often the failure data is not recorded. Also since the service personnel are usually not engineers, they often fails to identify different modes of failure.

All this points to a need for a more thorough and convenient instrumentation of elevator components. The current code calls for the installation of two earthquake protective devices for all traction elevators operating at a speed of 150 ft/min and above: (1) one seismic switch per building, and (2) one or more displacement switches on each counterweight.

The seismic switch, activated by the ground motion, is intended to forewarn that a potentially damaging earthquake is imminent. It is an acceleration-based switch, currently set at an acceleration threshold of 0.15g or lower. Whenever the acceleration intensity exceeds the threshold, set of actions are transmitted by the controller for the car to proceed in a pre-planned manner.

A displacement switch is a device that is actuated whenever the counterweight is abnormally displaced out of its guideway. A "ring-on-a string" device has been used for this purpose. It consists of one (or more) taut sensing wire that run along the hoistway, adjacent and parallel to the counterweight frame. The sensing wire is electrically charged. It passes through a ring attached to the counterweight frame. The unusual excursions of the counterweight beyond the normal limits of its operation bring the sensing wire in contact with the ring, which normally closes the open circuit to trigger a set of operational commands. The car is instructed to stop, then proceed slowly to the nearest landing in the direction away from the counterweight, open the door, and then stop finally.

There has been a mixed report about the utility of these seismic sensors that act as triggers. Onoda and his colleagues have developed a "seismic-wave energy" based forewarning system. It is not quite clear how this seismic wave energy measurement can be related to the vulnerability of elevator components such the guide rails, and roller guide systems. Suzuki and Kishimoto (1990) from Mitsubishi Corp. have also developed a forewarning system to detect the arrival of the P-and S-waves. The effectiveness of this forewarning system, however, has not been tested yet. It is not hard to believe that the ground acceleration measurements alone cannot predict the true dynamic state of a system like elevator. (See, Onoda and Ikeda, 1988; Onoda and Yamakoshi, 1990; Onoda et al, 1992).

From this discussion it is clear that more comprehensive sensor systems are required to gather data of engineering significance for technical assessment of these systems. The sensor are also needed for health monitoring to provided timely information that can be used for a robust and reliable control of the elevators in seismic emergencies. The information gathering system must be installable in the cramped elevator hoistways unobtrusively without any interference with moving part. The sensor system must recognize the special dynamic characteristics of the counterweight system to gather information optimally for maximum use. The feasibility of connecting different sensors should be carefully considered. The clutter of connecting wires hanging loose can often jeopardize the efficacy of a data collection system, especially during an earthquake. Fiber optic sensors offer several advantages, but multiplexing many (>6) of them still an open problem for research. Besides gathering data of scientific utility, the shared use of these sensing and processing system during an emergency situation is now a necessity, and it must be considered in designing these system.

ACKNOWLEDGEMENTS

This study is partially supported by a grant from Multi-Disciplinary Center for Earthquake Engineering Research at the State University of New York, Buffalo. This support is gratefully acknowledged.

REFERENCES

Onoda Y. and Ikeda E. (1988) Development of seismic wave energy sensing-type earthquake detector. *Proceedings of the Ninth World Conference on Earthquake Engineering, Tokyo-Kyoto, Japan*, VII: 661-666.

Onoda Y. and Yamakoshi T. (1990) Seismic-wave energy-sensing earthquake detector, in *Elevator World Educational Package and Reference Library*, 3: VI.12-VI.14.

Onoda Y., Nakazato M., Nara T. and Yokoi I. (1992) Development of elevator-use, popular-type seismic energy-sensing earthquake detector, *Proceedings of the Tenth World Conference on Earthquake Engineering, Madrid, Spain*, 10: 5959-5962.

Singh M. P., Suarez L. E. and Rildova (2002) Seismic Response of Rail Counterweight Systems of Elevators in Buildings, Technical Report, Submitted to Multidisciplinary Center for Earthquake Engineering Research, State University of New York, Buffalo.

Suzuki K. and Kishimoto F. (1990) An earthquake-emergency landing device with a primary-wave sensor for elevators, in *Elevator World Educational Package and Reference Library*, 3: VI.15-VI.16.

Efficacy of Optic fiber Strain Sensor for Determination of Internal Strain in 3-D Braided Composites

Shenfang Yuan [1], Rui Huang [2]

[1] *Department of Civil & Materials Engineering, University of Illinois at Chicago, Illinois, USA*

[2] *The Aeronautical Key Laboratory for Smart Materials and Structures, Nanjing University of Aeronautics and Astronautics, China*

ABSTRACT

In this article, a new method is introduced to study the mechanical performance of braided materials using co-braided fiber optic sensors. Experimental research is performed to devise a method of co-braiding the optical fiber into a 3-D structure. Effectiveness of this new testing method is evaluated. Experimental results show that multiple fiber optic sensors may potentially be braided into 3-D braided composites to measure parameters inside materials, providing a more accurate measurement method and leading to a better understanding of these materials.

1. INTRODUCTION

3-D braided composite technology has stimulated a great deal of interest in the world at large. The through-the-thickness, or interlacing, reinforcement in braided composites has the potential to eliminate or reduce the size of delamination and thus eliminate or reduce strength degradation due to accidental damage. In addition, braided composites will potentially cost less than traditional laminated composites.

A reliable understanding of the properties of 3-D braided composites is of primary importance for the successful utilization of these materials. However, methods for understanding and predicting the mechanical properties of braided composites tend to be more complex than those for laminated composites, as the braided yarns are not straight and the three-dimensional nature of these kinds of composites, coupled with the shortcomings of currently-adopted experimental test methods.

In this article, a measurement method is introduced which braids fiber optic sensors into 3-D braided composites to determine their internal parameters, such as strain, temperature, pressure and so on. Because of many particular advantages optic fiber sensors have, along with their ability to be easily adhered to composites, the embedding of fiber optic sensors in laminated composites to monitor damage, stress, strain and etc., has been broadly studied.

This paper reports on experimental research conducted to devise a method for braiding the fiber optic sensors into a 3-D structure. Effectiveness of this new testing method is evaluated on two counts. First, the optical performance of the fiber optic sensor is studied before and after braided into the 3-D braided composites, and also after the entire manufacturing process of the 3-D braided composites, in order to validate the ability of the optical fiber to withstand the manufacturing process. Second, two kinds of fiber optic sensors are co-braided into 3-D braided composites in order to compare their effectiveness in testing internal strain. One of these is the Fabry-Parrot (F-P) fiber optic sensor; the other is the polarimetric fiber optic sensor. Experiments are conducted to validate the performance of these sensors in testing strain under tension, bending and thermal environments in 3-D carbon fiber braided composite specimens, both locally and globally. Experimental results show that multiple fiber optic sensors may potentially be braided into 3-D braided composites to measure parameters inside materials, providing a more accurate measurement method and leading to a better understanding of these materials.

2. METHODS FOR CO-BRAIDING SENSORS

During manufacturing process, optical fibers are co-braided as axial yarn. They are just run straight and parallel with the specimen's braiding direction. Structure fibers are then braided around the optical fibers, as shown in Figure 1.

In this research, the performances of three kinds of typically used optical fibers, 1.3μm monomode optical fiber, 1.3μm and 1.55μm polarization maintaining optical fibers, are studied and compared. Girder form specimens are designed. The RTM molding method is employed. A special RTM molding die is designed to allow the optical fibers to exit the die without damage.

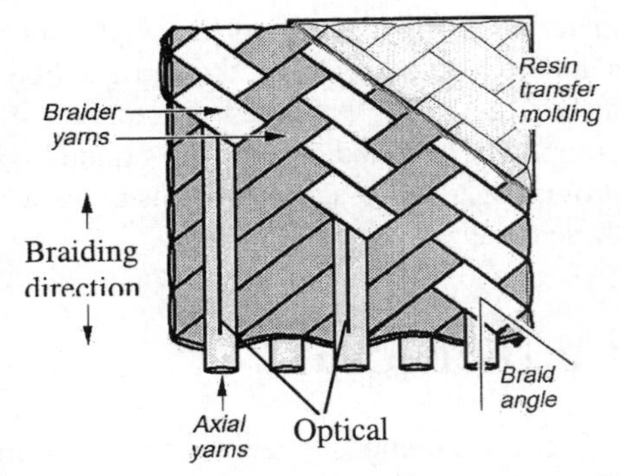

Figure 1. Optical fiber braided method

In the RTM process, the release agent is silicone oil, the resin is Epoxy 828, and the curative is polyamide. The temperature of the molding process is 85° C, and the molding model is under a 0.1MPa vacuum condition.

Made of 3k3 carbon fiber, the specimens are braided through a four-step method with specimen dimensions of 250 × 20 × 6 mm, a fiber volume content of 50 percent and yarn count of 28 × 8. Each piece of optical fiber co-braided in the braided specimens is two meters long, including the part in the specimen and the part outside of the specimen.

3. STUDIES OF OPTICAL PERFORMANCES

For the monomode optical fibers considered herein, three main performance parameters are chosen to be studied. They are cut-off wavelength, numerical aperture, and mode field diameter. In addition to these parameters, crosstalk – which reflects polarization ability – is considered for polarization maintaining optical fibers.

The measured performance parameters for these optical fibers are shown in Tables 1-3. Table 1 includes the main performance parameters of the optical fibers before braided into the braided composites. Table 2 contains the parameters of the optical fibers after braided, and Table 3 includes the parameters of the optical fibers after the RTM process.

Table 1. Main performance parameters of original optical fiber

Optical fiber	Cut-off wavelength (nm)	Mode field diameter (μm)	Numerical aperture	Crosstalk (db)
1.3μm monomode	1148-1119	7.24-7.25	0.156-0.158	/
1.3μm polarization maintaining	1096.1	6.84	0.181	-33.4
1.55μm polarization maintaining	1512.7	6.37	0.247	/

Table 2. Main performance parameters of optical fiber after co-braided into 3-D structure

Optical fiber	Cut-off wavelength(nm)	Mode field diameter (μm)	Numerical aperture	Crosstalk (db)
1.3-μm monomode	1133	7.27	0.155	/
1.3-μm polarization maintaining	1094	6.85	0.173	-35.7
1.55-μm polarization maintaining	1526	6.55	0.250	-33

Table 3. Main performance parameters of optical fiber after RTM process

Optical fiber	Cut-off wavelength(nm)	Mode field diameter (μm)	Numerical aperture	Crosstalk (db)
1.3-μm monomode	1060	7.27	0.150	/
1.3-μm polarization maintaining	1076	6.88	0.173	-33.9
1.55-μm polarization maintaining	1504	6.53	0.251	-29.6

Comparing Tables 1 and 2 shows that all optical performance parameters considered show little change after the fiber optic sensors have been braided into the 3-D composites using the proposed method. The differences may be attributed to inhomogeneity in the optical fiber introduced during the manufacture process. Comparing Tables 1 and 3, little change is noted in the mode field diameter and numerical aperture after the RTM process, which indicates that the ability of the optical fibers to receive light does not change substantially and the concentration grade of light in the optical fiber does not change appreciably after the RTM process. The cut-off wavelength exhibits a small decrease after the RTM process. In the RTM process, the optical fibers undergo high temperature and pressure, resulting in softened cladding of the optical fiber and leading to a change in optical performance, along with producing some residual stresses. The crosstalk value of the polarization maintaining optical fiber decreased after the RTM process. This indicates that the extent of the change of polarization state becomes smaller. In summary, then, the optical performance test results show that optical fibers can be braided in the 3-D braided composites safely and will survive the manufacturing process. The optical fiber performance characteristics do not change substantially during the manufacturing process.

4. EFFECTIVENESS ON INTERVAL STRAIN TEST

Two kinds of sensors are evaluated for their effectiveness in testing the internal strain of 3-D braided composites after braided into 3-D braided composite specimens. One is the extrinsic Fabry-Perot (F-P) interferometer; the other is the polarimetric fiber optic sensor. Two redundant sensors were embedded in each specimen to assure measurement capability in case of damage to one of the sensors during the manufacturing process. As to the specimens for the experiment using the F-P sensor, the position of the F-P cavity is braided at the center of the specimen to test the internal strain in a local area. The polarimetric fiber optic sensors can test the total strain along its length braided in the specimen, yielding a global strain measurement. Figure 2 shows the positions of the fiber optic sensors.

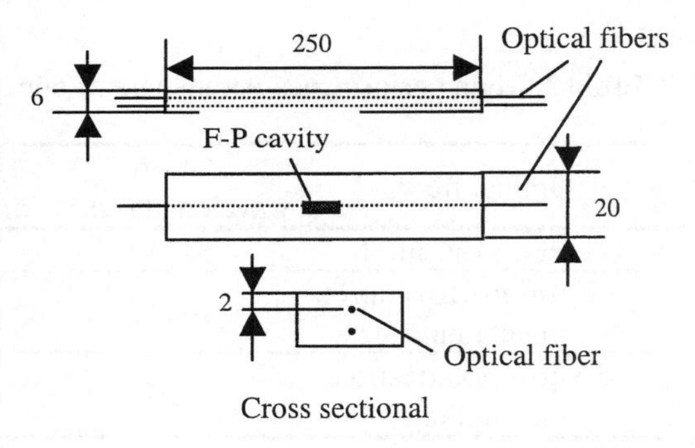

Figure 2. Experimental specimens

4.1. Tension strain measurement using F-P fiber optic sensor

In the experiment, a 60-kN mechanical testing machine is used. In order to validate the results measured by the fiber optic sensor, one strain gauge is also bonded to the specimen and a bridge circuit is adopted to measure the strain. In the case of this experiment, the surface strain and interval strain of the specimen caused by tension is the same, so surface mounted strain gauge can be used to evaluate the test result of optical fiber sensor.

The length scale for the F-P fiber optic sensor is 30mm. An HP83437 broadband light source having a center wavelength of 850 nm was used. Light output was analyzed using an HP86140A spectral analyzer. Strains measured both by strain gauges and F-P fiber optic sensors corresponding to a given load are compared in Figure 3.

The experimental results demonstrate that the strain-load relation tested by the braided F-P fiber optic sensor is quite similar to that tested by the strain gauge, indicating that the

F-P fiber optic sensor is effective in testing the local interval strain of 3-D braided composites after braided into the composites.

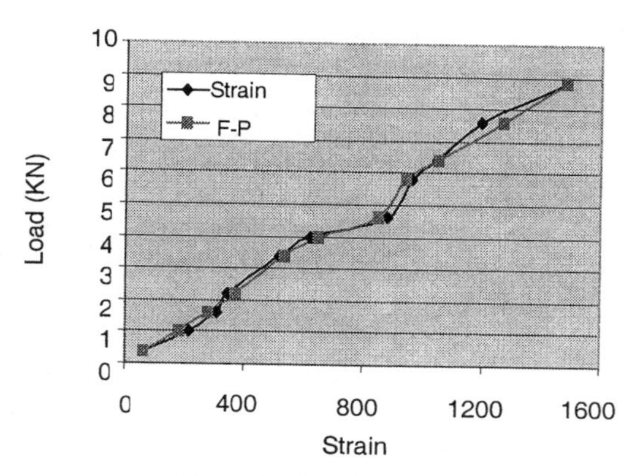

Figure3. Tensile test results

4.2. Bending specimen strain measurement

A three-point bending experimental system is adapted to test the internal strain under bending. Also, a strain gauge is attached to the specimen at the point below the applied load, and its output is compared to the measured results for the fiber optic sensor. In this case of experiment, although strain inside the specimen is different from that on the surface, but their relation can be calculated by mechanical analysis. So strain gauge measurement result still can be adopted to compare with that tested by optical fiber sensor.

The tested result transformed and compared in Figure 4. The x-axis represents the strain of the layer where the fiber optic sensor is located, and the y-axis represents the output of the polarimetric fiber optic sensor. The curve is similar to a sine wave, corresponding to the characteristic of a polarimetric sensor and demonstrating that the polarimetric fiber optic sensor is effective in testing global internal strain when braided into 3-D braided composites.

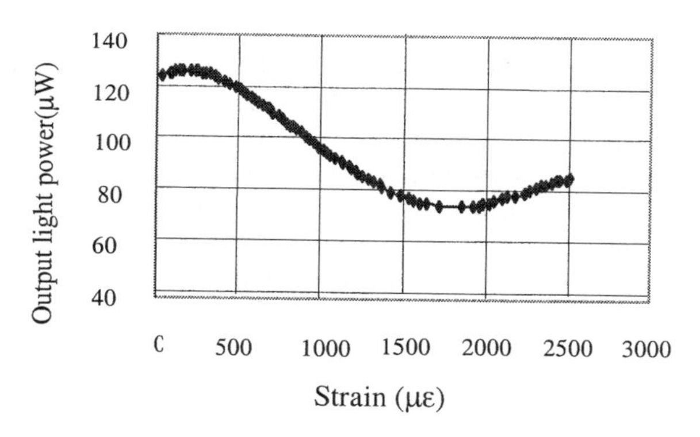

Figure 4. Bending strain test results

4.3. Thermal strain measurement using F-P sensor

The experimental specimen with an F-P fiber optic sensor braided in is put into an oven to test the thermal strains of the braided composite specimen at different temperatures. The long-term highest working temperature of the epoxy material is 120°C. Thus, the test temperature range is chosen from 20°C to 80°C and the thermal strain of the braided composite is measured in 5°C increments.

The experimental result is shown in Figure 5. It is known that 3-D braided composite has a negative thermal expansion performance in the braiding direction corresponding to an increase of temperature. In the case of carbon fiber epoxy braided composite materials, where the thermal strain of carbon fibers under 400°C is very small, the thermal expansion property is mainly controlled by resin type and is not linearly dependent on the temperature; some difference is observed in different temperature ranges. The experimental data shown in

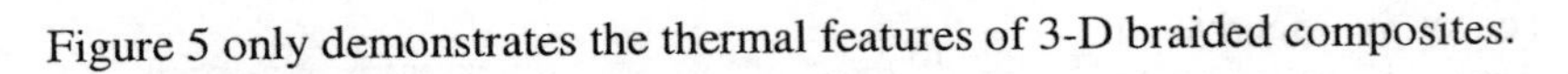
Figure 5 only demonstrates the thermal features of 3-D braided composites.

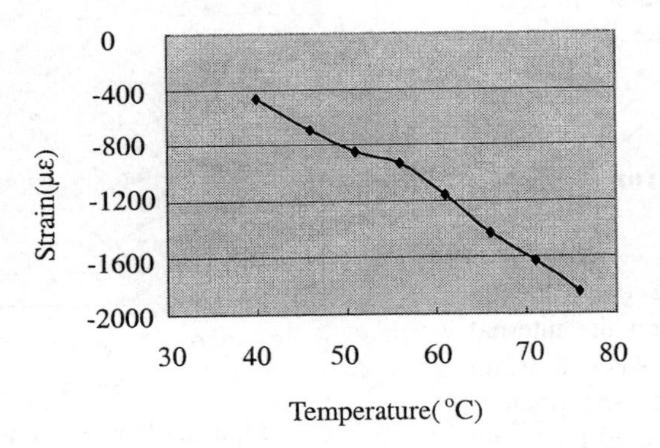

Figure 5. Thermal strain tested

5. CONCLUSIONS

Experimental results obtained through this research demonstrate that fiber optic sensors may potentially be braided into 3-D braided composites for the purpose of measuring parameters inside these materials. This approach can provide a more accurate measurement method by which to better understand 3-D braided composites. By braiding fiber optic sensors into 3-D braided composites, both surface and internal parameters can be tested – and on both local and global scales. In addition to its use in measuring the overall performance of 3-D braided composites, it is also envisioned that this method may be used to monitor the performance of the material during the RTM manufacturing process and throughout its service life.

References:

Ko FK. Braiding. In: Dostal CA, editor. Engineered Materials Handbook, vol. 1: Composites. ASM International; 1987: 519-528.

John E. Master, Marc A. Portanova, Standard Test Methods For Textile bComposites, NASA Contractors Reports 4751, 1996: 16-20.

Eric Udd. Fiber Optical Smart Structure. USA: Wiley-Interscience Publication, 1994: 1-21.

Baoqi Tao, Ke Xiong, Shenfang Yuan, et al, Smart Materials and Structures, The Defense Press of China, 1997: 45-67.

Liang Dakai, Huang Mingshuang, Tao Baoqi, Experimental Study on Optical Fiber Embedded in Carbon Fiber Composite Structure, Journal of Materials Engineering, Vol. 44 (2), 2000: 16-18.

Vikram Bhatia, Mark E.Jones, Jennifer L. Grace, *et al.* Applications of Absolute Fiber Optic sensors to Smart Materials and Structures. Proceedings of 10th Fiber Optic sensors Conference. 1994: 171-174.

Gui Yang, Daxin Ao, Zhiyong Zhang, *et al.* Manufacture, Technique and Industry Practice of Braided Composite Structure. China: Scientific Press, 1999: 19-67.

Low Frequency PVDF Sensors for Condition Monitoring of Structures

George M. Lloyd[1] and Ming L. Wang[2]

[1] *Postdoctoral Associate, Dept. of Civil and Materials Eng., University. of Illinois, Chicago, IL 60607.*
[2] Professor, Dept. of Civil and Materials Eng., University of Illinois, Chicago, IL 60607.

Abstract:

Polyvinylidene fluoride (PVDF) film is used to fabricate a new displacement gage. PVDF is active in nature, and it does not require any external electrical power for operation. This makes it an attractive choice for sensor arrays used for structure monitoring since it has substantially smaller power requirement and is much more cost-effective than resistance strain gages which are widely used now. A micropower charge amplifier is designed to condition the signal of the PVDF film. The complete PVDF sensor is designed around a charge amplifier and the PVDF film. The potentially detrimental pyroelectric characteristics of the PVDF film are eliminated through a temperature compensation scheme. To reduce noise from out-of-band EMI, a low-power filter is designed to filter the output of the sensor. The frequency response of the resulting hybrid PVDF sensor is measured using a random vibration method. The response is compared with simulations using a SPICE model. The results indicate that the high-pass frequency response of the PVDF gages matches concurs with the SPICE model. The low cutoff frequency can be reduced to values <0.1 Hz, making it suitable for structure monitoring.

1. Introduction

PVDF, whose molecular repeat formula is $(CH_2\text{-}CF_2)_n$, exhibits the strongest piezoelectric and pyroelectric activity of all known polymers (Luo 1999, Fukada 1981). With proper design, it can be used in transducers for sensing a number of different physical quantities. In monitoring and detecting damage in adhesive joints, the advantages of PVDF transducer over ceramic crystals are many. For example, PVDF is relatively inexpensive and the metallized films can easily be etched into intricate sensor patterns. Furthermore, the PVDF transducers respond to a wide-banded frequency (0 to 1GHz) and are heavily damped, advantageous features when dealing with thin structural materials. These lightweight and flexible devices can be bonded on structures permanently with minimal impact on the structural performance. Such sensors could measure various properties which influence joint strength, including degree of cure and the presence of porosity, damage, and debonding.

Thus far, PVDF sensors have been successfully used as shock gages and NDE transducers in pulse-echo and ultrasonic techniques to monitor curing and to detect porosity and crack propagation in different model joint geometries (Stiffer 1983, Fiorillo 1989 & 1992, Dutta 1990, Diprisco 1993, Obara 1995, Tang 1993). For the most part, however, the applications have been limited to the measurement of high frequency signals. This paper focuses on the design and verification scheme the authors employed in order to utilize PVDF films for a structure monitoring displacement gage. In particular, effort was placed on establishing the low frequency response of the sensor, which has not been adequately measured and modeled.

2. Design of the Sensor

2.1 Description of the Sensor Design

The basic design of the sensor is shown in Fig. 1. A pair of temperature-compensated PVDF films are attached to an arched metal plate; relative displacement of the ends is sensed

Fig.1 Photo of the PVDF sensor designed for low frequency infrastructure monitoring.

through changes in curvature which couple strongly to the d31 mode of operation of the PVDF film. The alignment of the PVDF film pair is shown in Fig. 2. The size of each film is 0.75*0.20inch.

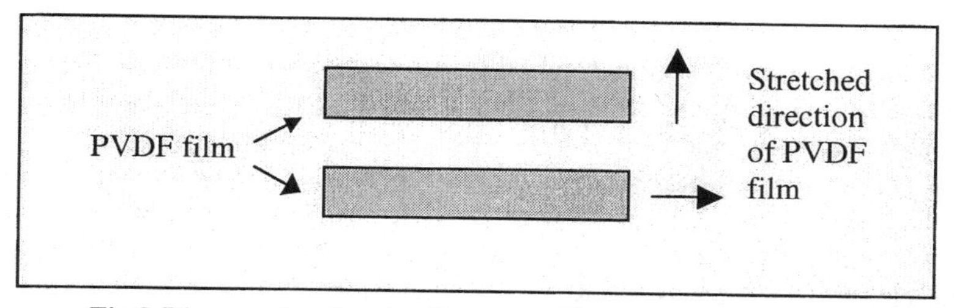

Fig.2 Diagram showing the alignment of PVDF film in the sensor

At low frequencies, the generated charge of deformed PVDF film is very small. The charge amplifier is needed to measure the output signal of the PVDF film under the mechanical disturbance, a type of operation termed voltage-mode sensing. The charge amplifier should be put close to the PVDF film, which is necessary to minimize the influence of stray electric fields which can cause low-frequency drifts. In order to permit voltage-mode operation and maintain the desirable characteristics of the PVDF films as a structure transducer, a micropower amplifier was designed to match PVDF film. A prototype of this amplifier is shown on the arch in Fig. 1. The circuit of the charge amplifier is shown in Fig. 3.

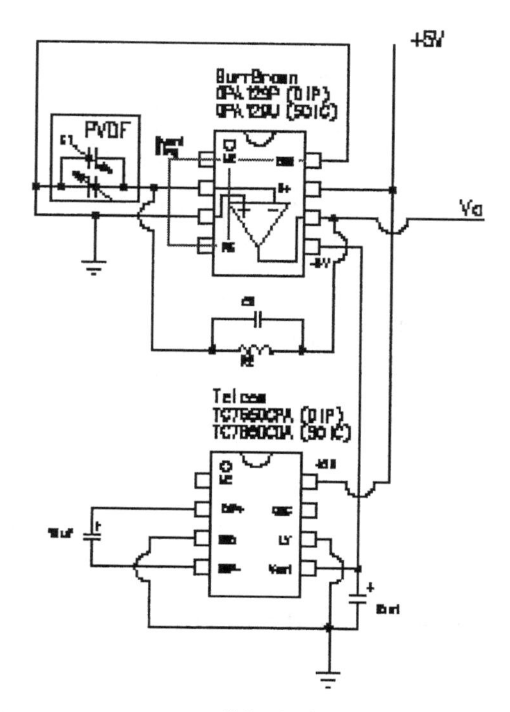

Fig.3 Schematic of micropower charge amplifier built around the Burr Brown OPA129

A Burr Brown OPA129 was selected because of its low input bias current, low voltage operation, and extremely low power consumption (microwatts to milliwatts). A Telcom DC/DC charge pump converter was used to provide negative bias to the charge amplifier, simplifying power requirements considerably. Even in the unoptimized prototype whose results we report here, the charge amplifier was small enough that it could be placed adjacent to the PVDF film. As shown later, the overall characteristics of this design provided high signal to noise ratio and eliminate integrator drift.

To reduce the disturbance of out-of-band noise, a filter was designed to filter the output of the sensor. Since a key non-structural noise component is 60Hz and most infrastructure excitations of interest lay below 25 Hz, the cutoff frequency of the filter was set at 45Hz.

A TLC04 Butterworth low-pass switched-capacitor filter was selected, which provided accurate fourth-order low-pass filter. A second Telcom DC/DC charge pump converter was also used to provide negative bias to the filter. The circuit of the filter is shown in Fig. 4. The resistance of the resistor connected to CLKR pin is 8.24Kohm. The capacitor in series with the resistor has 22.8nF capacitance. The cutoff frequency of the filter was confirmed via laboratory measurements to be close to 45Hz.

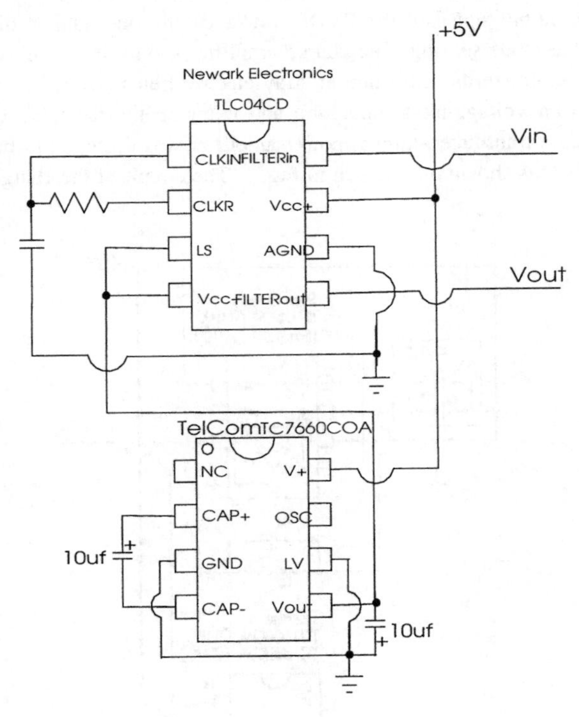

Fig.4 Schematic of Low-Pass Filter built around TLC04CD.

2.2 Temperature Compensation of the Sensor

PVDF film demonstrates a substantial pyroelectric effect. To eliminate this effect from introducing any influence on sensor repsonse, a bucking PVDF film was used to null out common-mode pyroelectric effects. (The configuration of the bucking gage is included in the schematic shown above in Fig. 2.) The length of the active sensor is along the stretch direction of the PVDF film. The dummy gage is cut along the direction perpendicular to the stretch direction. Importantly, the areas of both films are same. The positive electrode of the active gage is connected with the negative electrode of the dummy gage. The other two electrodes are connected together. The equivalent circuit is shown in Fig. 5.

Because the areas of active sensor and the dummy gage are same, the charge generated from the change of temperature are identical in magnitude. The pyroelectric component of any net charge will be eliminated since the opposite electrodes of two films are connected together, a result confirmed experiments discussed below.

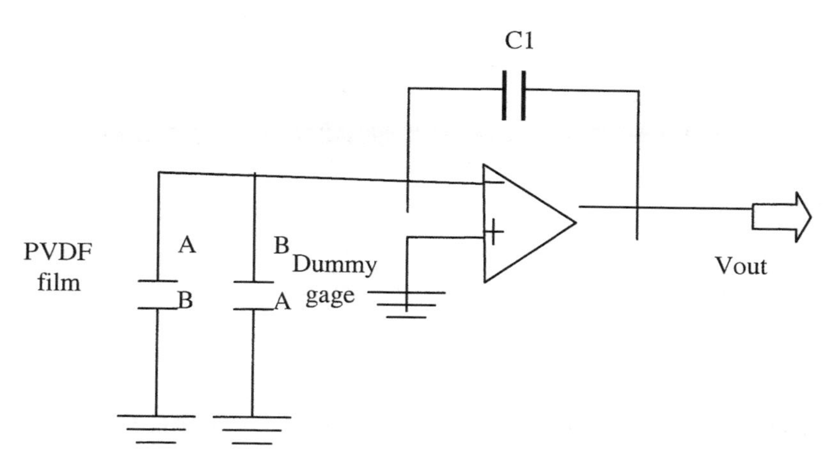

Fig.5 Equivalent circuit diagram for temperature compensation scheme used to null out pyroelectric effects.

A PVDF sensor with and without temperature compensation was moved from a room at 22 C to a constant-temperature oven at 60 C. The resulting output of two gages are shown in Fig. 6 and Fig. 7.

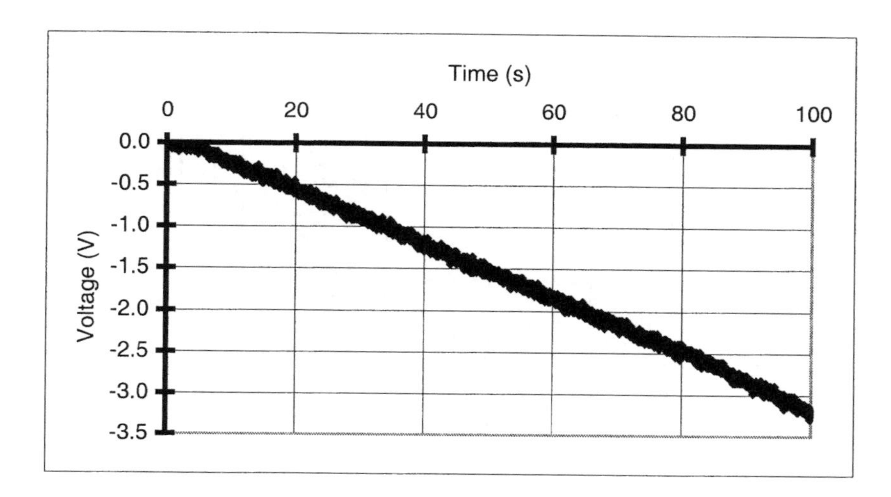

Fig.6 Pyroelectric effect of PVDF film

The output of the gage without temperature compensation clearly demonstrates the magnitude of the pyroelectric effect (Fig. 6). The output of the compensated gage, however, did not produce a measurable signal, as shown in Fig.7.

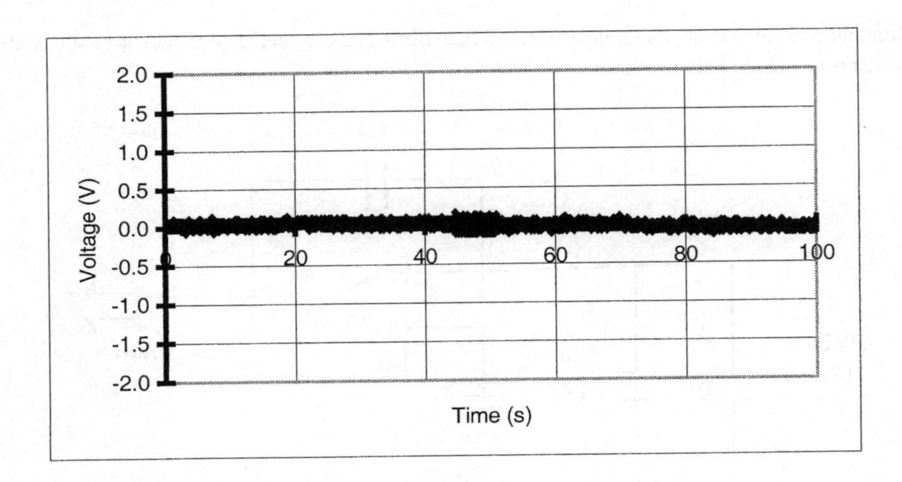

Fig.7 Output of PVDF sensor with temperature compensation while put in an increasing temperature condition

3. SPICE Model of the Sensor

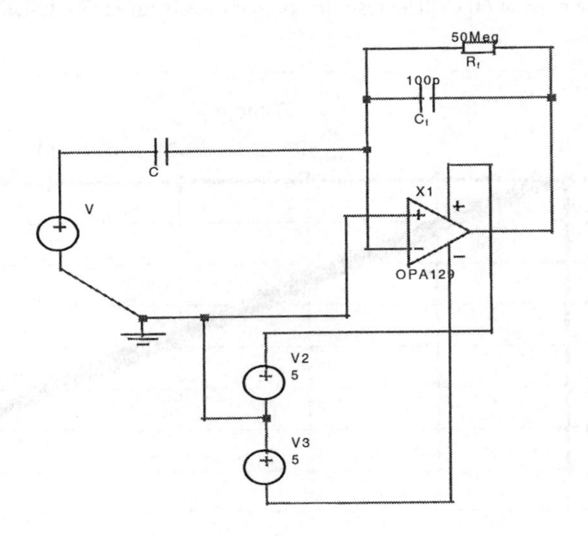

Fig.8 SPICE model of the PVDF sensor

The SPICE model of the sensor is shown in Fig. 8. The SPICE model for the charge amplifier was obtained from the manufacturer's WEB site. This model was derived and verified in reference (Zhang 2001).

4. Frequency Response of the Sensor

Characterization of the hybrid PVDF sensor frequency response was performed in a precision displacement stage using a random vibration method (Shoukens 1998). Stage actuation was via a VTS-40 electromechanical shaker with a DRC-1 random controller. The controller produced a continuous band-limited signal which was amplified by a linear amplifier and supplied to the displacement stage and sensor.

For the purpose of validation, the PVDF displacement sensor also contained a reference strain gage in addition to the PVDF strip. The strain gage, MEM CEA-B-125UW-350 (GF=2.115), was conditioned by a Vishay 3800 wide range strain indicator. The output of the indicator was digitized by one channel of a sampling scope. Records of $n=5*10^4$ points at 100 sps were collected. Analog prefiltering (cutoff frequency = 45Hz) was performed on both channels prior to digitization.

The reference strain gage used to measure relative motion of the stage was calibrated for both small and large displacements. The respective calibration factors for both ranges are in good agreement with one another, and consistent with a linear FEM analysis (Lloyd 2001).

In view of the difficulty of measuring the response at the low frequencies of interest and the stated floor of the DRC-1 controller, the feedback capacitor of the op-amp was varied such that the low-frequency response of the sensor could be compared with the SPICE prediction as a validation of the basic model. Three feedback capacitors (C_f) with values of 10pF, 56pF and 100nF were evaluated in the experiment. The feedback resistor (1Gohm) was manufactured by a specialty vendor (Micro-Ohm). Because of the dependence of the gain, r, on C_f for the range of values tested, post-gain scaling of the charge amplifier output was used to adjust the final recorded amplitude, as given in Table 1.

Table 1. Experiment Parameters

C_f	□seç	Post gain	r
100 (nF)	100	30 (dB)	0.0033
56 (pF)	0.056	0 (dB)	5.89
10 (pF)	0.010	-6.021 (dB)	33

Estimation of the frequency response of the sensor was performed by applying standard spectral density estimators to the unwindowed discrete fourier transform components (Lloyd 2001). The frequency response of the sensor and comparison with the SPICE simulator are shown in Fig. 9. The measured frequency responses of the sensor are matched very well with the SPICE simulator even though the sensor response shows a faster rolloff at low frequencies than predicated. The low cutoff frequency is reduced with the increase of the C_f value. When C_f =100nF, it is about 1Hz, which is suitable for many types structure monitoring applications, and it can be further reduced by appropriate selection of the passive feedback elements.

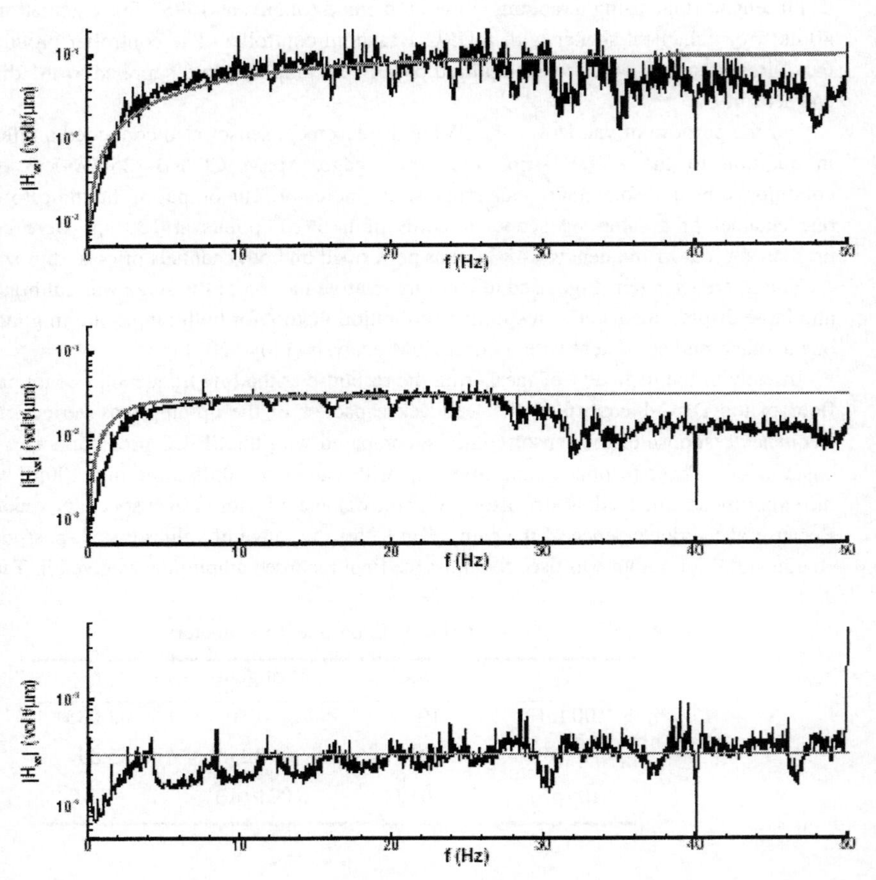

Fig.9 Measured frequency response of the sensor compared with SPICE model for several values of C_f (From top to bottom C_f = 10pF, 56pF, 100nF)

5. Conclusions

A charge-mode PVDF sensor was designed for structure monitoring. The measured frequency response of the sensor was compared with the simulation of the SPICE model. The sensor response shows a high-pass characteristic with low cutoff frequency about 1 Hz indicating that the design is suitable for structure monitoring.

References

1. DiPrisco, M. and Gandelli, A.(1993). "A new experimental approach to the investigation of contact forces at an interface." Materials and Structures. 26, 214-225.

2. **Dutta, Piyush K. and Kalafut, John(1990). "Evaluation of PVDF piezopolymer for use as a shock gage." Special report 90-23. U.S. Army Corps of Engineering: Cold Regions Research & Engineering Laboratory.**
3. **Fiorillo, A.S., Allotta, B., Dario, P. and Francesconi, R.(1989). An ultrasonic range sensor array for a robotic fingertip. Sensors and Actuators. Vol. 17, 103-106.**
4. **Fiorillo, A.S.(1992). "Design and characterization of a PVDF ultrasonic range sensor." IEEE Transations on Ultrasonics, Ferroelectrics and Frequency Control. 39(6), 688-692.**
5. **Fukada, E. and Furukawa, T.(1981). "Piezoelectricity and ferroelectricity in polyvinylidene fluoride." Ultrasonics. 19(1), 33-38.**
6. **Lloyd, George M., Zhang, Ying, Fabo, Peter, and Wang, Ming L.(2001). "Hybrid Frequency Response Characteristics of a Low-Frequency Charge-Mode PVDF Curvature Sensor Measured with a Random Vibration Method." ASME International Adaptive Structure and Materials System Symposium. Nov. 11-16, New York, New York.**
7. **Luo, H. and Hanagud, S.(1999). "PVDF film sensor and its applications in damage detection." J. Aerospace Eng. Jan. 1999, 23-30.**
8. **Obara, T., Bourne, N.K. and Mebar, Y.(1995). "The construction and calibration of and inexpensive PVDF stress gauge for fast pressure measurements." Shock Physics. Jan. 1995. 345-348.**
9. **Shoukens, J., Rolain, Y., and Pintelon, R.(1998). "Improved Frequency Response Function Measurements for Random Noise Excitations." IEEE Transactions on Instrumentation and Measurement. 47(1), 320-322.**
10. **Stiffer, R. and Henneke, H.E.G.(1983). "The application of Polyvinylidene Fluoride as an acoustic-transmission transducer for fibrous composite materials." Mat. E. 41, 956-957.**
11. **Tang, B., Mommaerts, J., Duncan, R.K., Duke, J.C., Jr. and Dillard, D.A.(1993). "Nondestructive evaluation of model adhesive joints by PVDF piezoelectric film sensors." Experimental Mechanics. June 1993. 102-109.**
12. **Zhang, Ying (2001). Master Thesis: "Design and Testing of a Polyvinylidene Fluoride Sensor For Monitoring of Structure." University of Illinois, Chicago.**

2. Dutta, P. [illegible] and [illegible] (19[illegible]) "Evaluation of PVDF [illegible]" Special report [illegible], U.S. Army Corps of Engineers, Cold Regions Research & Engineering Laboratory.
3. [illegible] (1999) [illegible] Films [illegible] Sensors and Actuators [illegible]
4. [illegible] (1992) "Design and characterization of [illegible]" IEEE Transactions on Ultrasonics, Ferroelectrics, and Frequency Control [illegible]
5. [illegible] (19[illegible]) "Piezoelectricity [illegible]" [illegible]
6. [illegible] Zhang [illegible] Wang, Ming [illegible] "Hybrid [illegible] Electrical Impedance [illegible]" [illegible] International Symposium [illegible] Smart Structures and Materials [illegible] New York.
7. [illegible] (199[illegible]) [illegible] applications in damage [illegible]
8. [illegible] (19[illegible]) [illegible]
9. [illegible] (1999) "Improved [illegible] Frequency Response Function Measurements [illegible]" IEEE Transactions on Instrumentation and Measurement [illegible]
10. [illegible] "The application of [illegible]" [illegible]
11. [illegible] evaluation of model adhesive joints [illegible] piezoelectric film [illegible]
12. [illegible] (2003) Master's Thesis [illegible] Design and Testing [illegible] University of Illinois, Chicago.

Competing Risks in Stochastic Modeling of Structure Deterioration

Grace L. Yang

Department of Mathematics
University of Maryland, College Park, MD 20742

INTRODUCTION

Structure deterioration is a random process for one cannot predict with certainty when a structure will change its condition and when it will fail. Such random deterioration can be studied by a non-stationary Markov process. In a Markov process, the change of a structure over time from one condition (or state) to another is described by a set of rates (or risks of change), $q_{ij}(t)$, for $i, j = 1, \ldots, k$, where k represents the total number of possible states of the structure. From these rates, the probability of change from one state to another at any time t can be derived.

Markov processes have been used extensively for monitoring disease progression in patients. Many of the existing methods for analyzing Markov processes were developed for such biomedical applications. The similarity between disease progression and structure deterioration suggests that Markov processes could be very useful for monitoring structural health and for studying interactions of different risk factors that contribute to structure deterioration.

We discuss statistical estimation of the competing risks, $q_{ij}(t)$, and some recently developed non-parametric estimation methods, with a special focus on the treatment of incomplete data. With immensely increased computing capability, the complexity of the formulas is no longer an issue for practical applications.

EXAMPLE

In statistics, the study of structure deterioration and damage detection fall in the domain of risk analysis. The analysis involves stochastic modeling of deterioration data and estimation of failure or survival probability of a structure. The deterioration of a structure typically undergoes several stages before it fails completely or is damaged. Therefore a description of failure probability should include staging. Suppose that a structure under investigation is rated according to one of the k possible states. We let $X(t)$ denote the state of the structure at time t. In risk analysis, the basic quantities are the transition probabilities that the structure will change from state i to j from time s to t, that is,

$$P_{i,j}(s,t) = P[X(t) = i|X(s) = j], \quad \text{for } s < t, \ i,j = 0,1,\ldots,k.$$

Based on appropriate data, we wish to find optimal methods for estimating these probabilities and derive other quantities of interest such as total life time or time to failure.

For illustration, let us consider the data collected on the bridges of New York City, see, e.g. Yanev and Chen (1997). The bridge structural condition, $X(t)$, is considered as a variable in time. The data consists of biennial and interim inspections of bridge condition. The bridge condition is rated from 1 to 8 computed on the basis of weighted averages of certain bridge component conditions as follows.

Rating 1 Potentially hazardous;

Rating 2 Used to shade between a rating of 1 and 3;

Rating 3 Serious deterioration or not functioning as originally designed;

Rating 4 Used to shade between a rating of 3 and 5;

Rating 5 Minor deterioration, but functioning as orgininally designed;

Rating 6 Newly repaired/maintenance performed;

Rating 7 New condition;

Rating 8 Unknown;

We have modified slightly the rating for illustration purposes. Here, most of the ratings are monotonically decreasing as the structure ages. Thus the best rating is given to 7 for the brand new condition and the worst is assigned to 1. Note, however, the highest rating is 8 which indicates that the bridge has no data during the inspection year, and it does not represent the best bridge condition. Note also that a repair/maintenance would reverse the rating from a lower to a higher rating, for instance, from 3 to 6 after repair. When the rates, $q_{ij}(t)$, are non stationary and the number of states $k \geq 3$, the derivation of the transition probabilities becomes difficult if the evolution of $X(t)$ is not monotonic.

SURVIVAL ANALYSIS and INCOMPLETE DATA

Ideal data sets are hard to come by. Data for risk analysis often contain censored values or partially observed values. A case in point is the above bridge example. A bridge may receive rating 8 during an inspection, which means no information. Incomplete data often poses a serious problem in statistical analysis. One cannot discard them without biasing the final analysis. A simple example illustrates this point. It may seem simple, but often it is not easy to compute the average survival time of a group of cancer patients from a medical follow-up study. This is because some patients may leave the study when they are alive. (There may be other reasons. But to keep the example simple, we shall not discuss them.) For these patients, who are loss to follow-up, it is only possible to record how long they have lived while in the program. Including them in the calculation would shorten the average survival time while excluding them would bias the findings and violate the principle of random sampling. Without random sampling, it would be difficult to generalize the findings. Partial measurements are not limited to medical data. Instrument limitations, such as dead time and sampling rate, can also result in incomplete measurements.

How to analyze incomplete data has been one of the main concerns of statisticians. The research on incomplete data dates back at least to the early 1950s with several major publications of which I will mention two. The first one is the publication of Fix and Neyman (1951) on "A simple stochastic model of recovery, relapse, death and loss of patients" in *Human Biology.* The authors used a 4-state homogeneous Markov chain to model the survival-ship of cancer patients. This is perhaps the earliest paper using Markov chain for staging survival time as well as addressing the issue of censored (missing) data. One of the four states is designated as the state of a patient losing to follow-up. Fix and Neyman employed a parametric model by assuming the risks of change, $q_{ij}(t)$, are constant in t. The second publication is by Kaplan and Meier (1958) who proposed a non parametric estimator for the survival probability with censored data. The method requires no parametric model assumption and is regarded as a breakthrough in statistics. These publications have generated tremendous interest in the statistics community and much progress has been made on the treatment of incomplete data since that time. I shall now turn to some of the current developments.

MARKOV MODEL and COMPETING RISKS

Many of the existing models for risk analysis can be viewed as special cases of a Markov process. These include the Fix-Neyman model (1951) and the Kaplan-Meier estimator (1958); although the approaches used in these two papers are very different.

For risk analysis, it is mostly a matter of preference whether one works with transition probabilities $P_{ij}(t)$ (defined above) or their derivatives or their integrals. Sometimes, the choice is for reasons of applications.

The risks (of change) are the derivatives of transition probabilities $P_{ij}(t,v)$ given by

$$q_{ij}(t) = \frac{dP_{ij}(t,v)}{dv}\,|_{v=t}\,. \tag{1}$$

We shall put the risks in a matrix form:

$$Q(t) = \begin{bmatrix} q_{11}(t) & q_{12}(t) & \dots & q_{1k}(t) \\ q_{21}(t) & q_{22}(t) & \dots & q_{2k}(t) \\ \vdots & \vdots & \vdots & \vdots \\ q_{k1}(t) & q_{k2}(t) & \dots & q_{kk}(t) \end{bmatrix}.$$

Depending on the field of applications, $q_{ij}(t)$ are variably called transition rates, intensities, risks, failure rates, forces of mortality, or hazard rates.

Since given any current state, say i, a higher probability of moving into the state j with rate $q_{ij}(t)$, implies a smaller probability of moving into other states, Neyman and Fix (1951) referred to them as *competing risks.*

In some applications *the duration of stay in each state* is a preferred quantity to risks. Then one uses the integrals

$$\int_0^t P_{ij}(0,v)dv.$$

The Fix-Neyman model has the number of states $k = 4$ and the risks of change, $q_{ij}(t)$, independent of time. The Kaplan-Meier estimator, on the other hand, corresponds to a three-state Markov chain ($k = 3$) with unspecified time dependent risks $q_{ij}(t)$.

STATISTICAL ANALYSIS

The basic data needed for estimation of the transition rates are the transition counts, $N_{ij}(t)$ which represents the number of *direct* transitions from state i to j in the time interval $(0,t]$. The collection $\{N_{ij}(t), i,j = 1,\dots,k, \text{ for } i \neq j\}$ is called a multivariate counting process. The data are dependent. The standard statistical methods do not apply.

The direct counting data are usually difficult to get. But if one has, then the transition rates $q_{ij}(t)$ can be estimated non parametrically. A major advantage of non parametric estimation is that it is free of any parametric assumption which is not alway easy to justify.

The counting processes approach to risk analysis is due to O. Aalen (1978). It unifies and generalizes many of the existing methods. Aalen used the modern theory of martingales and stochastic integrals to analyze the counting data and developed an0 (asymptotically) optimal estimation procedure (under certain assumptions) for the cumulative hazard rates $\beta_{ij}(t) = \int_0^t q_{ij}(t)$. The estimates are given by

$$\hat{\beta}_{ij}(t) = \int_0^t J_{ij}(s)(Y_{ij}(s))^{-1}dN_{ij}(s) \tag{2}$$

where $J_{ij}(s)$ is some function needed for adjusting censored observations and $Y_{ij}(t)$ are certain observation processes.

The estimates $\hat{\beta}_{ij}$ are unbiased and asymptotic normal. An estimate of the variance of $\hat{\beta}_{ij}$ is given by

$$\hat{\sigma}^2 = \int_0^t J_{ij}(s)(Y_{ij}(s))^{-2}dN_{ij}(s). \tag{3}$$

Aalen's approach has since become a standard method for risk calculations. Further treatment of Markov chains may be found in Aalen and Johansen (1978). There now exists a huge literature on the subject. For further developments and a historical note, see, e.g. Yang (1997).

CONCLUDING REMARKS

We gave a very brief presentation on the importance of non stationary Markov processes in multivariate risk analysis and in the treatment of incomplete data. It is a pleasure to acknowledge that certain Markov chains have been used for analyzing bridge deterioration. However these Markov chains are assumed to have monotonic sample paths. In other words, the matrix Q given below equation (1) is a triangular form. It would be interesting to consider a more general non-stationary Markov chain with non monotonic sample paths (for $X(t)$) so that missing values and maintenance can be accounted for in the analysis.

REFERENCES

[1] Aalen, O. O. (1978). *Non-parametric inference for a family of counting processes. Ann. Statist.* 6, 701-726.

[2] Aalen, O. O. and Johansen, S. (1978). *An empirical transition matrix for non-homogeneous Markov chains based on censored observations. Scand. J. Statist.* 5, 141-150.

[3] Fix, E. and Neyman, J. and (1951). *A simple stochastic model of recovery, relapse, death and loss of patients. Human Biology* 23, 205-241.

[4] Kaplan, E. L. and Meier, P. (1958). *Nonparametric estimation from incomplete observations.* J. Amer. Statist. Assoc. 53 457-481.
[5] Yang, G. (1997). *The Kaplan-Meier estimator.* Encyclopedia of Statistical Sciences, update vol. 1, Wiley, 334-343.
[6] Yanev, B. and Chen, X. (1997)*Life-cycle performance of New York city bridges, Transportation Research Record* 1389, 17-24.

Section 3

Implementation of New Technology in Inspection and Monitoring

Ta-Lun Yang

ENSCO, Inc.

INTRODUCTION

Technological advancements in measurement, communication and data management are providing new ways of inspection and monitoring in transportation safety. Traditional manual and periodical processes are being replaced by automated and continuous methods. Large quantities of data are analyzed real-time to extract crucial information. Concise and organized results are delivered to the appropriate personnel in a much timelier manner. More in-depth and historical information is readily available through high-speed on-line query. The new process of inspection and monitoring is forcing revolutionary changes in the system maintenance. Some real examples in the transportation industry are presented for illustration.

Two on-going monitoring and inspection programs in transportation are used to illustrate the many aspects of new technology being applied. All major track routes are inspected by special railcars equipped with computerized electronic measurement systems to insure compliance with respect to safety standards. Technologies deployed in two operational systems are described: the new high-speed passenger train Acela, operated by Amtrak between Washington, DC, and Boston, MA, and the National Automated Track Inspection Program operated by the Federal Railroad Administration. The employed technology includes non-contact sensing, high-speed real-time processing, differential global positioning, inertial dead-reckoning, satellite communication, wireless communication, palm computing, pager notification and web-based information dissemination. One can see from the illustrations that there are always plenty of new tools to treat old problems; gaining acceptance by the industry is generally a more challenging task.

THE NATIONAL AUTOMATED TRACK INSPECTION PROGRAM

The Federal Railroad Administration (FRA) of the Department of Transportation has been tasked by Congress to monitor the condition of all railroad tracks in the nation to insure compliance with the Track Safety Standards. The standards have been promulgated since 1973 and were revised substantially in 2000, primarily to cover the emerging high-speed passenger trains.

Track Geometry Measurements

The current inspection program employs a 100-ton, self-propelled rail vehicle that was custom designed and built for the mission (Figures 1 and 2). The state-of-the-art track measurement system uses an inertial platform and CCD-imaging technology to measure the

Figure 1 – New FRA Track Inspection Car T2000

Figure 2 – Drive Station of T2000 Track Inspection Car

geometrical characteristics of the track at the sample rate of once a foot (Figure 3). The measurements are compared with the Federal Standards. Each sample of geometrical measurements is "tagged" with location information, in terms of latitude and longitude, for geographical reference.

Location by DGPS and Report by Wireless Communication

Location and time information are obtained from a Differential Global Positioning System (**DGPS**) system that is capable of providing location information to the precision of better than one meter. The **differential correction** signal is obtained through **satellite communication** available as a commercial service.

The basic GPS and the differential correction information are available once a second. The location information is available with a delay of a fraction of a second due to decoding and processing. The concept of data flow is shown in Figure 4.

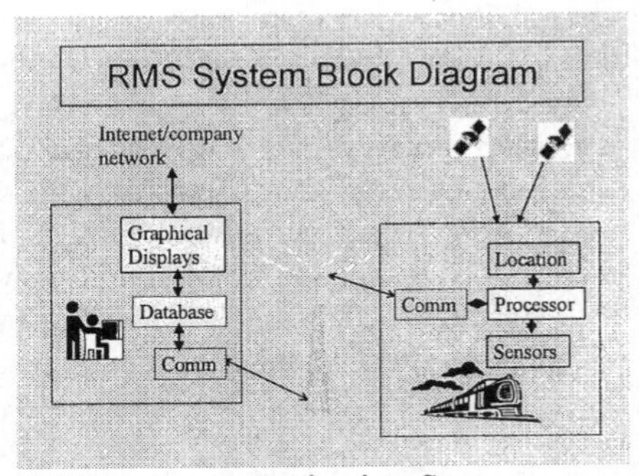

Figure 3 – Monitoring System Information Flow

Figure 4 – Truck-Mounted Optical and Inertial Sensor Package

Inertial Navigation

The inspection car would have moved over a hundred feet during this necessary but random delay. There are also stretches of track that are in tunnels or blocked by buildings or mountains from satellite visibility. The onboard measurement system makes use of the inertial sensors to perform **navigational dead reckoning** to fill-in the location information foot-by-foot when DGPS is not available for any reason (signal dropouts, tunnels, noise interference, etc.).

Application of Palm-Computing Technology

The users of the inspection data include some 150 Federal track inspectors, working in five regional offices, as well as engineers and track maintenance personnel of the railroad. The results of the inspection, including foot-by-foot track geometry data and analysis reports, are provided to the users in the form of **CD/ROM's** that are produced at the end of each test day. The data are entered into a national database that is available to authorized users over the web. A **Palm computer** is issued to each inspector. Inspection manuals, safety standards and forms are stored in the Palm unit for compiling and electronic filing of inspection reports (Figure 5). For data up-load, the forms are filled out by the inspector at the track site; the completed forms are electronically filed with headquarters through a desktop computer either via a serial-port cradle or an infrared port. For data download, exception lists generated by the inspection car are downloaded into the Palm computer. Each Palm computer is equipped with a **GPS attachment**. The navigation software in the Palm computer will guide the track inspector to each of the sites of a track geometry exception based on the location tag (longitude and latitude) attached to each detected defect. Track maintenance crews can find and repair each of the detected defects with the aid of the **GPS navigation feature**.

Figure 5 – Palm Computing Device

Offline Data Analysis and Web-accessible National Database

Offline software packages are available to review the data collected by the inspection car and compare the data between any two inspection surveys of the same track. Since distance sampling does not repeat perfectly from one inspection run to the next, data from the same track cannot be overlaid precisely due to a minute forward-and-back slippage of the distance base. In order to overcome this difficulty, a software algorithm was developed to make continuous and optimum adjustments of the distance scale on one set of data to achieve the optimum mathematical correlation with respect to another set of data that was collected over the same track at a different time. This software algorithm is named the **Automated Track Data Alignment System (ATDAS)**. Once aligned to the optimum correlation, the difference between the measured values can be used to reveal the slow but distinctive degradation of the track caused by traffic load and environmental inputs. Quantifying the rate of degradation allows track owners to forecast and plan maintenance actions to keep the quality of the track infrastructure adequate to meet the service demand.

Inspection results are stored in a **National Track Inspection Database**. Authorized users can logon to the web site and obtain new as well as historical information on every mile of track being covered by the national inspection program. Government inspectors have access to all the data; railroad operators are limited to the trackage owned by the railroad.

THE NEW AMTRAK HIGH-SPEED TRAIN ACELA

Amtrak inaugurated its new high-speed passenger service on the Northeast Corridor late 2001. The custom-built "Acela" (Figure 6) can reach a top speed of 150 mph as opposed to the 125-mph top speed for the traditional "Metroliner" trains. In addition to the absolute liner speed, all trains are subject to curving speeds that are defined through maximum cant deficiency (or unbalanced elevation) to limit the centrifugal force on vehicles and passengers. Due to the use of body-tilting technology, higher cant-deficiency (thus higher curving speed) is allowed for the Acela. Passenger comfort is maintained in the same way as banking of airplanes in turns.

Figure 6 – Amtrak 150-mph Acela Locomotive

The Federal Track Safety Standards was substantially modified and expanded to accommodate the introduction of Acela and other planned high-speed services. The class of track has increased from six to nine, addressing high-speed service up to 200 mph. Inspection standards and inspection frequency have been updated, and the requirement for monitoring ride quality was introduced for high-speed passenger services. The net result is the reduction of total travel time between Washington, DC and Boston, MA from 7 hours and 35 minutes to 6 hours and 33 minutes.

Track Geometry Inspection

Amtrak has two track inspection cars, designated 10002 and 10003, that cover the Northeast Corridor high-speed track and all the passenger routes in the country. Inspection car 10003 is embedded in one of the Acela trainsets; it is operated whenever the trainset goes into service. Both of the inspection cars are equipped with measurement instruments similar to those used on the FRA inspection car T2000 described earlier. Amtrak's Northeast Corridor track route has been surveyed by a "Fly Map" technology. Aerial video scanning technology was combined with DGPS technology to create precise longitude/latitude coordinates for every feature on the track route. The coordinate information was loaded into the onboard system, and the onboard DGPS system compared the measured train location with the look-up table and made **automatic location marks** of pre-selected targets (mileposts, bridges etc.) in the data stream.

Catenary Geometry Measurement

In addition to track measurement instrumentation, car 10002 is also equipped with catenary measurement equipment. It uses **fiber-optic sensors** (Figure 7) to measure the lateral position of the overhead wire and a pantograph height measurement system to determine wire height. Measurements are made dynamically under life-wire conditions at regular train speed. The measured positions are converted to track-based coordinates. Out-of-spec conditions are identified and reported for corrective action. Video recordings are made of the dynamic pantograph-catenary interaction along with the measurements.

Figure 7 – Fiber Optic Sensor Array for Catenary Lateral Position

Autonomous Ride Quality Monitoring

At the present time, 12 **Remote Monitor Systems (RMS)** are installed on selected Acela cars for monitoring ride quality. Each RMS (Figure 8) is an unmanned autonomous system that measures and analyzes ride quality according to specifications. Location, speed and time are obtained from a GPS system included in the RMS. A wireless communication system sends the processed results regularly to a central station. The central station receives reports from all twelve remote systems and displays summarized results (Figure 9) on a **Geographic-based Information System (GIS)** (Figure 10). If an event is of particular interest, the central-station operator can request more detailed information from the remote system (Figure 11).

Figure 8 – RMS Sensor, GPS and Communication Package

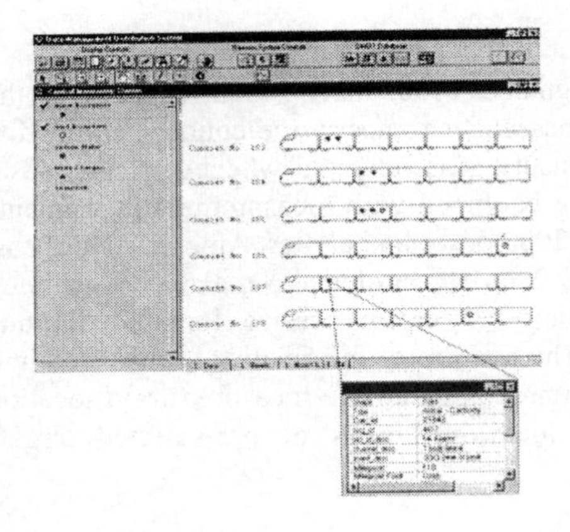

Figure 9 – Central-Station Display (Summary by Vehicle)

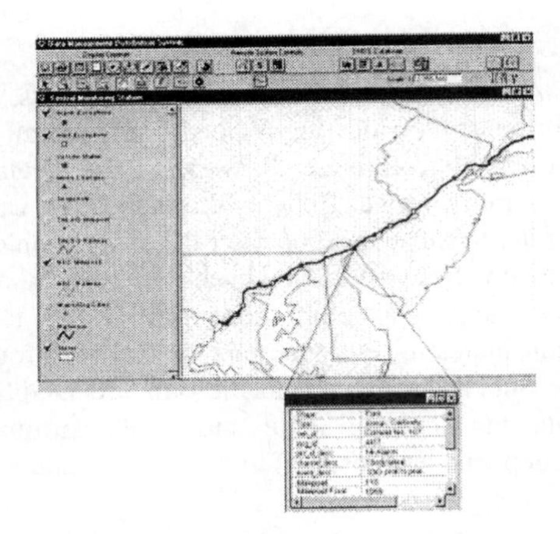

Figure 10 – Central-Station Display (Summary by Location)

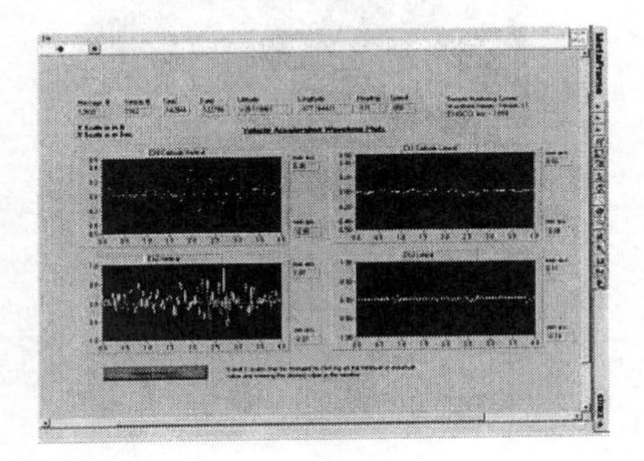

Figure 11 – Detailed Data on Demand

Measurement of Dynamic Wheel-rail Interaction Forces

Amtrak inspection cars are equipped with instrumented wheelset to measure wheel-rail interaction forces. Standard wheels (Figure 12) are installed with strain gauges to measure the stresses at selected locations on the wheel plate. Software programs are developed to convert the measured stresses to continuous vertical and lateral force measurements at the wheel-rail contact point. These dynamic forces are the key physical parameters that are related to wear and potential for derailment.

Figure 12 – Instrumented Wheelset

Satellite Images for Infrastructure Management

Fabio Casciati

Department of Structural Mechanics, University of Pavia, Italy

ABSTRACT

Satellite images represent a mature technology that for different reasons is delayed in becoming of public use.
This paper, after a brief summary of the technology main aspects, lists the most important reasons wich prevent from an indiscriminate use and emphasizes the potential of its applications in structural engineering.

1. INTRODUCTION

Having in mind structural and infrastructural engineering applications, one put himself/herself in an area different from the exploitation of satellite image technology within scientific applications. There, historical words as LANDSAT or SPOT denote the first steps and SAR and its variations are the current trend. In terms of resolution (the actual size of the pixel) one moved from 80m (LANDAT) to 10m (SPOT) or to the range of SAR applications (from 18m to 7m). No matter that this resolution is not suitable for civil engineering applications.
The first investigation in the area (Casciati and Giorgi, 1996a; Casciati and Giorgi, 1996b; Casciati et al.,1996; Casciati, 1997; Casciati et al., 1997a; Casciati et al., 1997b) moved to a different technology emerging from the hidden paths of military secrets. The Russian images with 2m of resolution (figures 1 and 2) were the first adopted. The market was waiting for the exploitation of the military archives with resolution spanning in the range from 0.5m to 0.1m, or in the launch of an initiative (by Lokhed) promising images with 1m-0.5m resolution.
One had to wait more than 5 years before the IKONOS images (figure 3) became available.

Figure 1: Details of satellite image representing the downtown of Catania (sensor KVR-1000, resolution 2m), as distributed (a) and with suitable zoom (b).

Figure 2: Details of the Umbria Region of 1997 Earthquake as distributed (a) and with suitable zoom(b)

2. PREVENTING DIFFUSION

After the so-called Gulf War, the technology of low-resolution satellite images, mainly developed within the military apparatus of countries as Russia, USA, UK, France and Israel, appeared mature for its exploitation in the civil world. Almost ten years later one realizes that new countries (as China and India) are making formidable progresses in the area, but in the meantime the situation in the countries first developing the technology is nearly frozen.
The authors can only guess some of the hidden causes, while some deep rational also comes from the rights that the modern democracy tends to assign the citizen.

Among the hidden causes one sees:

1) the decision of the military apparatus to maintain such images in the class of classified material; in this situation it is difficult to allows private companies to assemble such a delicate information;
2) it was the law, years ago, to pass any image taken from aircrafts or helicopters to the national Air Force authority which was allowed to filter (the idea was to cancel, but one better says to emphasise) the images from military installations. A development of satellite images for civil applications would lead to a direct communication from the satellite owner to the user.

The reasons related with social habit are:

1) when moving below the 50 cm resolution, the privacy of the citizen can be formidably violated;
2) the lawyers are discussing today the framework of law around the internet facility; they are not yet ready to discuss the code update which a diffusion of image exploitation would require. It is useful to remind here that the embryo of such a discussion is on the way when the images are collected by cameras located in public areas (street crosses, exterior of banks and public offices and so on).

The main tool for preventing diffusion is to delay launches which would guarantee a fixed period between images. The China projects report 8 hours of gap. Geo-stationary satellite would also guarantee a better coverage.

3. IMAGES ELABORATION

Once the satellite image is made available and images of the same resolution are expected to be collected in the future, one has to conceive a system architecture which consists of the following steps:

1) geo-referentiation of the image, so that the integration with existing databases is made possible;
2) database integration, by which each pixel in the image can be associated with the available information on civil infrastructures, critical buildings, monuments and any structural compnent which shoul be monitored;
3) edge detection, by which one identifies the current state of single buildings and structural compónents;

comparison between edges detected from images collected at different times: this emphasises the progress of landslides, the evolution of a river flood, the distribution of a fire as well as the damage showed by buildings.

Figure 3: satellite image representing the town of Pavia (Italy) as it appears at delivery; sensor IKONOS, resolution 1 m.

Figure 4: zoom on "Ponte vecchio" in Pavia

4. CONCLUSION

This paper proposes once again the support of satellite technologies in the structural assessment of large buildings and important components of infrastructure systems.
After their promising introduction in the middle of the nineties, marketing, rather than technological reasons prevented from their fast growth.
Today, with the IKONOS approach is time again to see satellite images as a strongly innovative support to structural monitoring and control.

ACKNOWLEDGEMENTS

The paper was supported by a grant from the Italian National Research Council (CNR), contract number 01- 01019-PF42.

REFERENCES

F. Casciati and F. Giorgi (1996a). GIS for Earthquake Damage Assessment, *Proc. of the first International Conference on Computing and Information Technologies for Architecture, Engineering and Construction*, Singapore, 16-17 May 1996.

F. Casciati and F. Giorgi (1996b). A Telematic System for Seismic Damage Assessment, *Proc. of the First International Conference on Structural Control*, Barcelona, Spain; published by World Scientific, Singapore, 165-171.

F. Casciati, L. Faravelli and F. Giorgi (1996c). Satellite Infrastructure Monitoring, *Proc. 2nd International Symposium on Civil Infrastructure Systems*, Hong Kong, 110-115.

F. Casciati, L. Faravelli (1996). Analisi costi benefici di intervento sul patrimonio monumentale, *Proc La protezione del patrimonio culturale. La questione sismica*, Roma

F.Casciati (1997). La vulnerabilita' sismica del patrimonio culturale. Aspetti informatici (in Italian), *Atti del II Seminario di Studio La Protezione del Patrimonio Culturale - La Questione Sismica*, Roma, 69-79

F. Casciati, P. Gamba, F. Giorgi, A. Marazzi and A. Mecocci (1997a). A Flexible Environment for EarthquakeRrapid Damage Detection and Assessment, *Proc. of the 1997 IEEE Geosci. and Remote Sensing Symp.*, Singapore, Vol.I, 113-115.

F. Casciati, P. Gamba, F. Giorgi and A. Mecocci (1997b). A Seismic Damage Detection Telematic Tool, *Proc. Unweltinformatik'97*, Strasburgo, 832-841.

F. Casciati, L. Faravelli and F. Giorgi (1998). Satellite Image GIS for Rapid Damage Assessment, in Aatre V.K., Varadan V.K. and Varadan V.V. (eds.) *Smart Materials,Structures and MEMS*, Proceedings of SPIE, vol. 3321, 0277-786X/98, 116-125.

M.J. Carlotto (1997). Detection and Analysis of Change in Remotely Sensed Imagery with Application to Wide Area Surveillance, *IEEE Trans. on Image Proc.*, Vol.6, No.1, 189-202.

L. Faravelli L, F. Giorgi and G. Zonta (1998). Infrastructure Components of Architectural Value: Prioritization and GIS Potential, in F. Casciati, F. Maceri, M.P. Singh and P. Spanos (eds.), *Civil Infrastructure Systems: Intelligent Renewal*, World Scientific, Singapore,

P. Gamba and F. Casciati (1998), GIS and Image Understanding for Near Real-Time Earthquake Damage Assessment, *Photogrammetric Engineering and Remote Sensing*

F. Casciati L. Faravelli and M. Pagani (1999). Seismic Damage Assessment by Image-Analysis and GIS Technologies, Proc. Petra '99 EU Workshop, Petra, Jordan

Dynamic Piezoelectric Shape Control Applied to Shells of Revolution with Translatory Support Excitation

Hans Irschik , Michael Krommer, Manfred Nader and Uwe Pichler

Division of Technical Mechanics, Johannes Kepler University of Linz, A-4040 Linz, Austria

ABSTRACT

Dynamic shape control of structures by piezoelectric actuation is considered in the context of vibrations produced by a translatory imposed support motion. The shape control problem under consideration is to identify a transient piezoelectric actuator distribution which, when superimposed upon the support excitation, produces a vanishing total deformation of the structure. Hence we wish to influence the motion of the structure such that it performs a purely translatory rigid-body motion in case the piezoelectric actuation acts together with the imposed support motion. An exact three-dimensional solution of this shape control problem is presented. Subsequently, the solution is applied to thin shells of revolution.

1. INTRODUCTION

The present paper deals with dynamic shape control of structures by piezoelectric actuation. We study transient disturbances of a structure produced by a translatory imposed support motion. The shape control problem under consideration is to identify a transient piezoelectric actuator distribution which, when superimposed upon the support excitation, produces a vanishing total deformation of the structure. Hence we wish to influence the motion of the structure such that it performs a purely translatory rigid-body motion in case the piezoelectric actuation acts together with the imposed support motion. In order to solve this problem, we assume the support motion to be known, and to be synchronous, i.e. to be instantaneously the same at every point of the supported boundary of the structure. We consider the case of small strains, linear piezoelastic constitutive relations and isothermal conditions.

Piezoelectricity has been frequently used in the literature for disturbance sensing and control of flexible smart structures, see Rao and Sunnar (1994), (1999), and Tzou (1998). The group of the present authors has contributed to this field with respect to both, static as well as dynamic shape control problems of smart beams and trusses, see Irschik et al.

(1998), (1999), (2001) and Nader et al. (2002). For a further literature review on shape control, see Irschik (2002). So far, our group dealt with shape control of structural vibrations by imposed forces. The present paper gives an extension with respect to translatory support motions. We first present a three-dimensional solution, which we apply to the case of thin shells of revolution afterwards.

In the subsequent derivations, we treat the deformation and stress due to the support motion as a fictitious body force problem, since the deformation due to a synchronous translatory support motion is analogous to a time-dependent self-weight of the structure. We talk about the piezoelectric problem when referring to the superimposed piezoelectric actuation. The three-dimensional solution of the dynamic shape control problem stated in the present paper represents an extension of a formulation by Irschik and Pichler (2001) for shape control of force-induced vibrations by means of thermal actuation. The solution of Irschik and Pichler (2001) rests the theorem of work expended and on Graffi´s theorem, see Gurtin (1972) for details on these theorems. Using anisotropic thermoelastic constitutive equations, the two theorems were connected to the force problem and to the thermal actuation problem by means of a dummy force problem. From the resulting convolution statements it was found by Irschik and Pichler (2001) that, in order to make the total displacement zero everywhere in the body, the thermal actuation stress must be the quasi-static stress induced in the force problem, minus the stress induced by the effect of thermal coupling in this force problem. Subsequently, this result is analogously formulated in the framework of the linear theory of piezoelasticity, i.e. in terms of piezoelectric actuation stresses. Particulary, we discuss the case of a translatory support motion, which is formulated as a fictitious body force problem (a time-dependent self-weight). The thermal actuation stress is replaced by the piezoelectric actuation stress, the latter corresponding to the inverse piezoelectric effect. Thermal coupling is replaced by the direct piezoelectric effect. In the present paper, this coupling effect is neglected. It is shown that the solution of the shape control problem is considerably simplified in the present case of a time-dependent self-weight, i.e. for a synchronous translatory support motion. The corresponding three-dimensional solution of the dynamic shape control problem is then applied to vibrations of thin shells of revolution with a translatory support motion and with piezoelectric actuators. The predicted solution is validated by means of dynamic Finite Element computations. These numerical results give an excellent evidence for the correctness of our analytical solution of the above shape control problem.

2. SHAPE CONTROL BY PIEZOPELECTRIC EXCITATION

The present Section presents some results derived by Irschik and Pichler (2001) for dynamic shape control of force induced vibrations by means of actuating thermal expansion strains, and which have been reformulated in order to be applicable to the case of piezoelectric actuation by Irschik and Pichler in (2001). These results rest upon two fundamental theorems of the linear theory of continuum mechanics. One of these is the theorem of work expended, see Gurtin, (2001), Sect.18. The theorem of work expended refers to a body B which is in equilibrium under the action of a system of body forces $\mathbf{b}_{(f)}$ and surface tractions $\mathbf{t}_{(f)}$. Let this system of forces produce a static admissible stress $\hat{\mathbf{S}}_{(f)}$, such that $\mathbf{t}_{(f)} = \hat{\mathbf{S}}_{(f)}\mathbf{n}$ at every regular point of the surface ∂B of B. The outer unit normal vector to

∂B is denoted by $\mathbf{n}$. The theorem of work expended connects the set $[\mathbf{b}_{(f)}, \hat{\mathbf{S}}_{(f)}]$ to any admissible deformation field $\tilde{\mathbf{u}}$ with a linear strain $\tilde{\mathbf{E}} = \frac{1}{2}(\mathrm{grad}\tilde{\mathbf{u}} + (\mathrm{grad}\tilde{\mathbf{u}})^T)$. Any solution of a physically meaningful equilibrium problem with given boundary conditions of traction or place represents an admissible stress field $\hat{\mathbf{S}}_{(f)}$. In the following, we refer to a boundary value problem, in which the surface tractions $\mathbf{t}_{(f)}$ are prescribed on the part ∂B_2 of $\partial B = \partial B_1 \cup \partial B_2$, while the corresponding displacements $\mathbf{u}_{(f)}$ are prescribed on the remaining part ∂B_1.

For further use, note that the forces $\mathbf{b}_{(f)}$ and $\mathbf{t}_{(f)}$ in our force problem may be transient, likewise to the displacements $\mathbf{u}_{(f)}$ prescribed at ∂B_1. However, only the quasi-static part of stress, $\hat{\mathbf{S}}_{(f)}$ must be used in the theorem of work expended. Note, that a hat refers to a quasi-static quantity in the following. By the notion of a quasi-static quantity we mean a solution in which the influence of the inertia forces has been omitted.

We particularly consider admissible deformation fields $\tilde{\mathbf{u}}$, which are due to a set of dummy forces $\tilde{\mathbf{b}}$ and $\tilde{\mathbf{t}}$. The dummy forces are imagined to be prescribed within B and on certain parts of ∂B, respectively. We assume that the dummy tractions $\tilde{\mathbf{t}}$ are prescribed on the same part ∂B_2 of ∂B, on which the tractions $\mathbf{t}_{(f)}$ have been prescribed in the force problem. The admissible deformation field moreover shall vanish on that part ∂B_1 of the boundary, on which the deformations have been prescribed in the force problem, such that $\tilde{\mathbf{u}} = \mathbf{0}$ on ∂B_1. In order to point out that we deal with this restricted class of admissible deformation fields, the field $\mathbf{u}$ will be denoted as a dummy deformation field subsequently. We emphasize that the dummy forces need not to be constant in time, and hence in general they will give raise to some inertia forces.

We then consider the system of time-dependent forces at a certain time τ, namely $\mathbf{b}_{(f)}(\tau)$ and $\mathbf{t}_{(f)}(\tau)$. The corresponding quasi-static stress is $\hat{\mathbf{S}}_{(f)}(\tau)$. Furthermore, we take the dummy fields at a shifted instant of time, $t-\tau$. We let $0 \le \tau \le t$, and we integrate the theorem of work expended from 0 to t. Adapting the convolution notation introduced in Sect. 10 of Gurtin (1972), we eventually arrive at

$$\int_{\partial B_2} \mathbf{t}_{(f)} * \tilde{\mathbf{u}}\, dA + \int_B \mathbf{b}_{(f)} * \tilde{\mathbf{u}}\, dV = \int_B \hat{\mathbf{S}}_{(f)} * \tilde{\mathbf{E}}\, dV \ . \tag{1}$$

In a second step we consider piezoelectric actuation problems. Again, the displacements are assumed to be prescribed on ∂B_1, while the surface tractions have to vanish on ∂B_2. The piezoelectric actuation may vary with the time, as well as the prescribed surface displacements. This problem will be called the piezoelectric problem in the following. The convolution version of the theorem of work expended then gives the following orthogonality relation between the dummy strains and the quasi-static part $\hat{\mathbf{S}}_{(p)}$ of the piezoelectrically induced stress,

$$\int_B \hat{\mathbf{S}}_{(p)} * \tilde{\mathbf{E}}\, dV = 0 \ . \tag{2}$$

Recall that $\hat{\mathbf{S}}_{(p)}$ denotes the quasi-static stress due to piezoelectric activity in which the influence of the inertia forces has been neglected.

As a second theorem crucial for the following derivations, we utilize Graffi's dynamic reciprocal theorem of the linear theory of elasticity, see Gurtin (1972), Sect. 19 and Sect. 61. Particularly, we apply Graffi's theorem to the force problem, the piezoelectric problem, and the dummy force problem introduced above.

Let $\mathbf{u}_{(f)}$ be the dynamic displacement, let $\mathbf{S}_{(f)}$ be the dynamic stress, and $\mathbf{E}_{(f)}$ the dynamic strain associated with the force problem. By the notion of a dynamic quantity we mean a solution of the problem in which the influence of inertia forces has been taken into account. Let furthermore the dynamic dummy stress be $\tilde{\mathbf{S}}$. Then, by the convolution version of the theorem of power expended, Eq.(1), Graffi's theorem can be re-written as

$$\int_{\partial B} \tilde{\mathbf{t}} * \mathbf{u}_{(f)}\, dA + \int_B \tilde{\mathbf{b}} * \mathbf{u}_{(f)}\, dV - \int_B \tilde{\mathbf{S}} * \mathbf{E}_{(f)}\, dV$$

$$+ \int_B \rho\, (\tilde{\mathbf{u}}(0) \cdot \dot{\mathbf{u}}_{(f)} + \tilde{\mathbf{v}}(0) \cdot \mathbf{u}_{(f)})\, dV$$

$$= \int_B (\hat{\mathbf{S}}_{(f)} - \mathbf{S}_{(f)}) * \tilde{\mathbf{E}}\, dV + \int_B \rho\, (\mathbf{u}_{(f)}(0) \cdot \dot{\tilde{\mathbf{u}}} + \mathbf{v}_{(f)}(0) \cdot \tilde{\mathbf{u}})\, dV\,. \tag{3}$$

The mass density of the body is denoted by ρ. The initial displacement and velocity of the dummy force problem are $\tilde{\mathbf{u}}(0)$ and $\tilde{\mathbf{v}}(0)$, respectively, while $\mathbf{u}_{(f)}(0)$ and $\mathbf{v}_{(f)}(0)$ denote the initial values of the force problem.

We achieve a further re-formulation by means of the linear anisotropic piezoelastic constitutive equations for the force problem and the dummy force problem,

$$\mathbf{S}_{(f)} = \mathbf{C}[\mathbf{E}_{(f)}] + \mathbf{d}[\boldsymbol{\varepsilon}_{(f)}] \tag{4}$$

and

$$\tilde{\mathbf{S}} = \mathbf{C}[\tilde{\mathbf{E}}] + \mathbf{d}[\tilde{\boldsymbol{\varepsilon}}]\,, \tag{5}$$

respectively. The fourth order elasticity tensor is denoted by $\mathbf{C}$, and the third order tensor $\mathbf{d}$ is the tensor of piezoelectric coefficients. The electric field vector associated with the force problem is $\boldsymbol{\varepsilon}_{(f)}$, while $\tilde{\boldsymbol{\varepsilon}}$ denotes the electric field in the dummy force problem. The quantities $\mathbf{d}[\boldsymbol{\varepsilon}_{(f)}]$ and $\mathbf{d}[\tilde{\boldsymbol{\varepsilon}}]$ are stresses induced by the inverse piezoelectric effect.

Taking into account the symmetry of $\mathbf{C}$, see Gurtin (1972), Sect. 28, we eventually obtain the form

$$\int_{\partial B} \tilde{\mathbf{t}} * \mathbf{u}_{(f)}\, dA + \int_B \tilde{\mathbf{b}} * \mathbf{u}_{(f)}\, dV - \int_B \mathbf{d}[\tilde{\boldsymbol{\varepsilon}}] * \mathbf{E}_{(f)}\, dV = \int_B (\hat{\mathbf{S}}_{(f)} - \mathbf{d}[\boldsymbol{\varepsilon}_{(f)}]) * \tilde{\mathbf{E}}\, dV$$

$$+ \int_B \rho(\mathbf{u}_{(f)}(0) \cdot \dot{\tilde{\mathbf{u}}} - \tilde{\mathbf{u}}(0) \cdot \dot{\mathbf{u}}_{(f)} + \mathbf{v}_{(f)}(0) \cdot \tilde{\mathbf{u}} - \tilde{\mathbf{v}}(0) \cdot \mathbf{u}_{(f)})\, dV\,. \tag{6}$$

In a second step, we run through the above procedure for the piezoelectric problem. The linear constitutive relations for this problem are

$$\mathbf{S}_{(p)} = \mathbf{C}[\mathbf{E}_{(p)}] + \mathbf{d}[\boldsymbol{\varepsilon}_{(p)}] \,, \tag{7}$$

where $\boldsymbol{\varepsilon}_{(p)}$ is the electric field produced in the piezoelectric problem. The dynamic displacement due to piezoelectric actuation is denoted by $\mathbf{u}_{(p)}$, and $\mathbf{E}_{(p)}$ is the corresponding dynamic strain. The term $\mathbf{d}[\boldsymbol{\varepsilon}_{(p)}]$ is the piezoelectric actuation stress. We finally arrive at the relation

$$\int_{\partial B} \tilde{\mathbf{t}} * \mathbf{u}_{(p)}\, dA + \int_B \tilde{\mathbf{b}} * \mathbf{u}_{(p)}\, dV - \int_B \mathbf{d}[\tilde{\boldsymbol{\varepsilon}}] * \mathbf{E}_{(p)}\, dV = -\int_B \mathbf{d}[\boldsymbol{\varepsilon}_{(p)}] * \tilde{\mathbf{E}}\, dV$$

$$+ \int_B \rho\, (\mathbf{u}_{(p)}(0) \cdot \dot{\tilde{\mathbf{u}}} - \tilde{\mathbf{u}}(0) \cdot \dot{\mathbf{u}}_{(p)} + \mathbf{v}_{(p)}(0) \cdot \tilde{\mathbf{u}} - \tilde{\mathbf{v}}(0) \cdot \mathbf{u}_{(p)})\, dV \,. \tag{8}$$

We now choose an uncoupled dummy problem, $\mathbf{d}[\tilde{\boldsymbol{\varepsilon}}] = \mathbf{0}$, with zero initial conditions, $\tilde{\mathbf{u}}(0) = \mathbf{0}$, $\tilde{\mathbf{v}}(0) = \mathbf{0}$. Adding Eq.(6) and Eq.(8) then yields

$$\int_{\partial B} \tilde{\mathbf{t}} * (\mathbf{u}_{(f)} + \mathbf{u}_{(p)})\, dA + \int_B \tilde{\mathbf{b}} * (\mathbf{u}_{(f)} + \mathbf{u}_{(p)})\, dV$$

$$= \int_B (\hat{\mathbf{S}}_{(f)} - \mathbf{d}[\boldsymbol{\varepsilon}_{(f)} + \boldsymbol{\varepsilon}_{(p)}]) * \tilde{\mathbf{E}}\, dV$$

$$+ \int_B \rho((\mathbf{u}_{(f)}(0) + \mathbf{u}_{(p)}(0)) \cdot \dot{\tilde{\mathbf{u}}} + (\mathbf{v}_{(f)}(0) + \mathbf{v}_{(p)}(0)) \cdot \tilde{\mathbf{u}})\, dV \,. \tag{9}$$

Since the dummy forces are arbitrary in Eq. (9), we arrive at the following theorem for the total displacement of the force problem and of the piezoeleastic problem: When the piezoelectric actuation stress is the quasi-static stress in the force problem (minus the stress induced by the indirect piezoelectric effect),

$$\mathbf{d}[\boldsymbol{\varepsilon}_{(p)}] = \hat{\mathbf{S}} - \mathbf{d}[\boldsymbol{\varepsilon}_{(f)}] \,, \tag{10}$$

and when the initial conditions and the kinematical boundary conditions are equal but opposite in sign for the two problems, $\mathbf{u}_{(f)}(0) = -\mathbf{u}_{(p)}(0)$, $\mathbf{v}_{(f)}(0) = -\mathbf{v}_{(p)}(0)$, and $\mathbf{u}_{(f)} = -\mathbf{u}_{(p)}$ at ∂B_1, then the total displacement vanishes everywhere within the body and at every time-instant,

$$\mathbf{u}_{(f)} + \mathbf{u}_{(p)} = \mathbf{0} \,. \tag{11}$$

This theorem is the desired solution of the dynamic shape control problem, see Irschik and Pichler (2001). If we neglect the indirect piezoelectric effect, Eq.(10) takes on the simple form

$$\mathbf{d}[\boldsymbol{\varepsilon}_{(p)}] = \hat{\mathbf{S}}_{(f)} . \tag{12}$$

In order to make the total displacement zero everywhere in the body, the piezoelectric actuation stress should be quasi-static stress in the force problem.

We note that the appearance of a quasi-static stress in Eq.(12) facilitates a practical application of the above solution of the dynamic shape control problem. When the force loading is separable in space and time, and when we consider the uncoupled theory, the quasi-static stress is also separable in space and time, and the distributed piezoelectric actuation then can be tailored for the specific force problem under consideration.

3. IMPOSED SUPPORT EXCITATION

The above results now are applied to the case of an synchronous translatory support excitation. Let this support excitation be $\mathbf{u}_{(supp)}(t)$, which is prescribed at ∂B_1. We denote the relative displacement of the structure due to the support motion as $\mathbf{u}_{(f)}$, such that $\mathbf{u}_{(f)} = \mathbf{0}$ at ∂B_1. The shape-control problem to be considered is as follows: We wish to superimpose a piezoelectric actuation with $\mathbf{u}_{(p)} = \mathbf{0}$ at ∂B_1, such that the structure moves as a rigid body with total displacement $\mathbf{u}_{(supp)}(t)$, i.e. without any deformation relative to the support excitation. In order to solve this shape control problem, it is recalled that the relative displacement $\mathbf{u}_{(f)}$ due to the synchronous translatory support excitation can be computed by considering the structure to be fixed at the supports, $\mathbf{u}_{(f)} = \mathbf{0}$ at ∂B_1, and to be loaded by a fictitious body force of amount

$$\mathbf{b}_{(f)} = -\rho \frac{d^2}{dt^2} \mathbf{u}_{(supp)} . \tag{13}$$

Hence, when the piezoelectric actuation stress is equal to the quasi-static stress produced by the fictitious body-forces of Eq.(13), then the total relative displacement will vanish, $\mathbf{u}_{(f)} + \mathbf{u}_{(p)} = \mathbf{0}$, compare Eq.(12). Note that, since $\mathbf{u}_{(supp)}(t)$ is space-wise constant, the quasi-static part of stress can be easily computed from the static stress due to the self-weight of the structure. The self-weight is to be applied in the (fixed) direction of the translatory support motion $\mathbf{u}_{(supp)}(t)$, and the acceleration of the free fall is to be set to unity. According to Eq.(13), the latter static stress then has to be multiplied by the support acceleration, $d^2\mathbf{u}_{(supp)} / dt^2$, in order to obtain the quasi-static stress required in Eq.(12). Since the spatial distribution of the quasi-static stress does not change in the present case, we can work with a fixed spatial distribution of piezoelectric actuators as well, where the time-evolution is to be chosen according to $d^2\mathbf{u}_{(supp)} / dt^2$. This of course facilitates the practical application of our concept.

4. APPLICATION TO THIN SHELLS OF REVOLUTION

In a co-operation with Siemens Cooperate Technology MS2 Munich (Prof. H. Meixner, Dr. H.-G. v. Garssen) we have applied the above concept to thin shells of revolution. A typical shell of the class under consideration is shown in Fig.1 The shell is subjected to a synchronous translatory support motion of its upper boundary, see Fig.1. The other boundary of the shell is free.

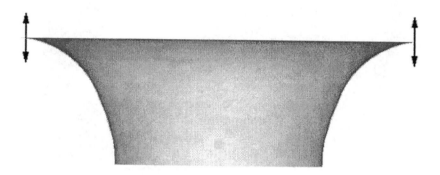

Figure 1.

The shell has been modelled by means of the Finite Element code ANSYS. The quasi-static stress due to the self weight of the structure has been computed, and the spatial distribution of the piezoelectric actuators has been evaluated by writing Eq.(12) in the framework of the theory of thin shells of revolution. This will be reported in detail elsewhere. The obtained spatial distribution of the piezoelectric actuation has been approximated by three ring actuators, each of them consisting of two layers attached at the outer and inner surface of the shell, see Fig. 2. The ring actuators are subjected to a time-dependent electric voltage fitting to the support excitation according to Eqs. (12) and (13).

Fig. 3 - 5 show the transient vibrations at a typical location of the shell relative to a sinusoidal support excitation. Shown are components of the relative displacement due to the support motion, $\mathbf{u}_{(f)}$, the displacement due to the piezoelectric ring actuators, $\mathbf{u}_{(p)}$, and the total displacement relative to the support excitation. Fig.3 is for a excitation slightly below the first eigenfrequency, Fig.4 shows the typical behaviour in case of a resonant excitation, and Fig.5 is for an excitation slightly above the first eigenfrequency. Since we work with the ring actuators, an exact cancellation of the relative displacement can not be achieved. The results in Figs. 3 - 5 nevertheless give a satisfactory evidence for the validity of the proposed theoretical concept of shape control even in the case of a resonant excitation.

Figure 2.

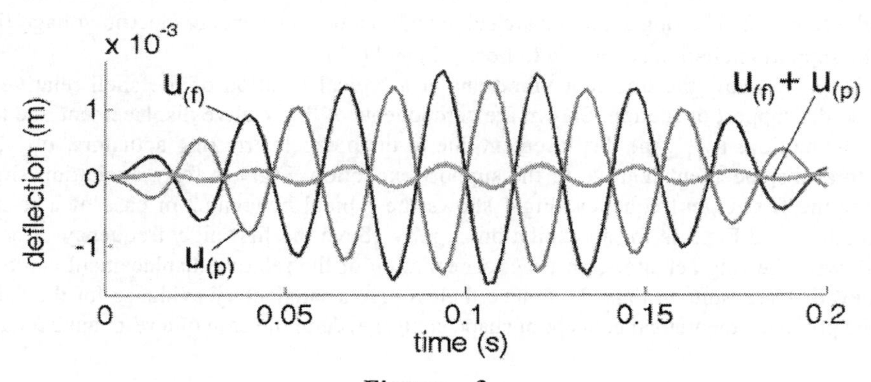

Figure 3.

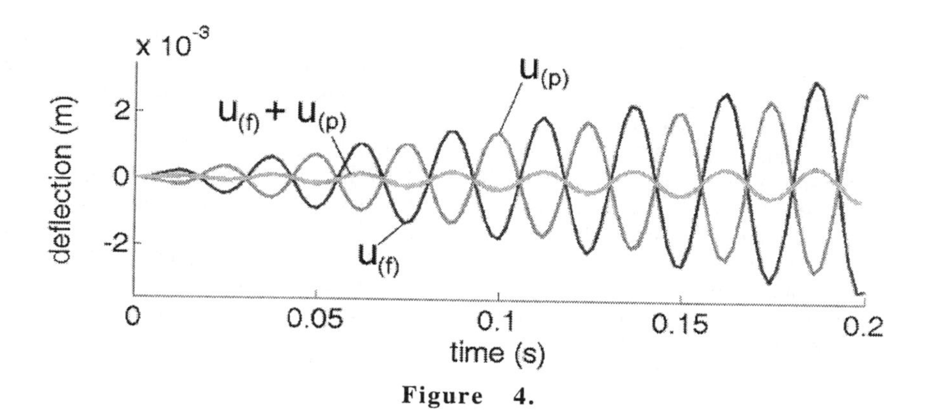

Figure 4.

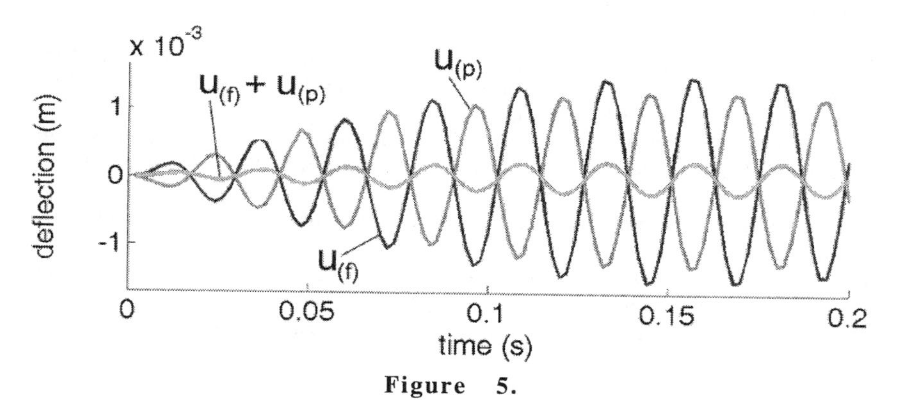

Figure 5.

ACKNOWLEDGEMENT

Support of the present contribution in the framework of the Austrian K+, "Linz Center of Competence in Mechatronics" (LCM), is gratefully acknowledged. The paper especially refers to projects 2.8 and 4.4 of LCM. The authors are grateful to Siemens Cooperate Technology MS2 Munich(Prof. H. Meixner and Dr. H.-G. v. Garssen) for their co-operation. and support

REFERENCES

Gurtin, M. E.. (1972) The Linear Theory of Elasticity, in S. Flügge (ed.), *Handbuch der Physik*, Vol. VIa/2, C. Truesdell (ed.), *Festkörpermechanik II*, 1-296, Springer-Verlag, Berlin.

Hagenauer, K., Irschik, H. and Ziegler, F. (1997) An Exact Solution for Structural Shape Control by Piezoelectric Actuation, in U. Gabbert (ed.), Smart Mechanical Systems - Adaptronics, vol. 244, VDI-Fortschrittsberichte, Reihe 11, Schwingungstechnik, pp. 93-98, Düsseldorf.

Irschik, H. (2002) A review on static and dynamic shape control of structures by piezoelectric actuation, *Engineering Structures*, 24:5-11.

Irschik, H, Heuer, R., Adam, Ch. and Ziegler, F. (1998) Exact Solutions for Static Shape Control by Piezoelectric Actuation, in Y.A. Bahei-El-Din and G.J. Dvorak (eds.), Proc. IUTAM Symposium on Transformation Problems in Composite and Active Materials, pp. 247-258, Kluwer, Dordrecht.

Irschik, H., Krommer, M., Belyaev, A. K. and Schlacher, K. (1999) Shaping of Piezoelectric Sensors/Actuators for Vibrations of Slender Beams: Coupled Theory and Inappropriate Shape Functions, *Int. J. Intelligent Material Systems and Structures*, 9:46-554.

Irschik, H.., Krommer, M. and Pichler, U. (1999) Annihilation of Beam Vibrations by Shaped Piezoelectric Actuators: Coupled Theory, CD-ROM Proc. 137th regular meeting of the Acoustical Society of America, 2nd convention of the EAA: Forum Acusticum – integrating the 25th German Acoustics DAGA Conference, Berlin, Germany.

Irschik, H., Krommer, M. and Pichler, U. (2001) Collocative Control of Beam Vibrations with Piezoelectric Self-Sensing Layers, in U. Gabbert and H. S. Tzou (eds.), *Proc. IUTAM-Symposium on Smart Structures and Structronic Systems*, Magdeburg, 2000, 315-322, Kluwer, Dordrecht.

Irschik, H. and Pichler, U. (2001) Dynamic Shape Control of Solids and Structures by Thermal Expansion Strains, J. Thermal Stresses, vol. 24, pp. 565-576.

Irschik, H. and Pichler, U. (2001) Shape Control of Structural Vibrations by Distributed Actuation. Proceedings of the European Meeting on Intelligent Strucutres, EMIS 2001 (A. Baratta, O. Corbi, eds.) Ischia, Italy.

Nader M., Gattringer H., Krommer M., Irschik H. (2002) Shape Control of Flexural Vibrations of Circular Plates by Piezoelectric Actuation, to appear in: ASME *Journal of Vibration & Acoustics*.

Rao, S. S. and Sunar, M. (1994) Piezoelectricity and Its Use in Disturbance Sensing and Control of Flexible Structures, *Applied Mechanics Reviews*, 47:113-123.

Rao, S. S. and Sunar, M. (1999) Recent Advances in Sensing and Control of Flexible Structures Via Piezoelectric Material Technology, *Appl. Mech. Rev.*, 52:1-16.

Tzou, H. S. (1998) Multi-field Transducers, Devices, Mechatronic Systems, and Structronic Systems with Smart Materials, *The Shock and Vibration Digest*, 30:282-294.

Electro-Magneto-Acoustic Transducers for Monitoring and Health Assessment of Metal Structures

Rahmat A. Shoureshi and Sun Wook Lim

Colorado School of Mines (CSM), Power Engineering Research Center, Golden, Colorado

ABSTRACT

A key component of a smart structure is its sensory system. Sensors provide necessary information about the state of the structure, its diagnosis and health assessment. *Sensing System* is an emerging topic of national and international interest in a variety of fields from living systems, e.g. biological, to non-living, e.g. engineered systems, to energy systems and infrastructures. Convergence of sensor technologies, communications, and computing has the potential to overcome barriers of time, scale, materials and environment. This paper presents the development of an innovative sensor technology and informatics for monitoring and health assessment of metal structures. Analysis design and implementation results of this sensor are presented.

INTRODUCTION

In order to eliminate or at least minimize discomforts and losses caused by man-made and natural phenomena, the concept of 'smart structures' has been proposed to mitigate the risk associated with the operation of physical infrastructure. A *smart structure* involves point or distributed sensors and actuators, and one or more processing units that analyze the responses from the sensors and uses advance signal processing and control methods to command actuators to apply localized strains that achieve some performance objective. Thus, it is envisioned that a smart structure has the capability to self-diagnose and adapt to a changing environment (such as loads and shape changes) as well as to a changing internal environment (such as damage or failure) these characteristics require advanced sensor technologies.

Extensive effort has been devoted to the development of sensors for infrastructures. In this paper, we have focused on those structures that have metal frames either in solid form or wire rope composition. Thus the new sensor is designed for this class of structures. Wire ropes form a major component of some of the key civil infrastructures, as well as part of the driving mechanism for a myriad of industrial products. Electric utilities and cable companies provide the backbone infrastructure for residential, commercial and industrial sectors. The cable structures mentioned above all suffer from various forms of degradation; including wear, natural corrosion, fatigue (due to wind and ice loading or undamped vibrations), surface embrittlement, and accidental damage. Furthermore, such structures are usually the critical component of a system that would involve significant property damage and/or loss of life, if the structure were to fail. Health assessment of these particular structures, is therefore, critical from a public safety point of view.

In the mining industry, the current practice for assessing the safety of wire rope [9] used in mineshaft hoists is to section off a 3-meter length of the wire rope and perform a variety of visual inspections and mechanical tests. Testing by this method is known to be unreliable, and not at all representative of the reliability and integrity of the remaining portion of the wire rope still being utilized [3,7,14]. In other applications, such as OHL, destructive testing is usually not a viable option, so non-destructive evaluation (NDE) is the only cost-effective method for health assessment. A variety of NDE techniques have been applied to assess the overall integrity of wire ropes and steel frames. In a series of recent journal articles from a trade publication specifically devoted to the wire rope industry [8-12], a number of different methods are compared to one another, highlighting the necessity of developing newer methods and/or a deeper understanding of existing methods in order to successfully perform NDE on wire ropes and steel frames.

The goal of any NDE measurements of a metal structure is to identify defects that have developed while in service. These are typically divided into two categories: loss of cross-sectional area, due to corrosion or damage, and localized faults (broken or missing strands). Visual inspection methods would obviously be ineffective for detecting either of these defects until they have reached the surface of the cable.

A majority of the current NDE work in structures is performed using acoustic emission (AE), as developed earlier in [1,2,5,6,13]. AE has been shown to be quite sensitive to wire breakage, and can warn of impending failure of a wire rope when the rope is loaded to approximately one-half of the maximum safe load [2]. AE has also been shown to provide information concerning wire or frame breakage during fatigue tests [10]. The AE technique seems to be well suited for health monitoring of wire ropes in situations where the transducers can be in pace at all times, and the wire rope is subjected to sufficient load to initiate the acoustic events recorded by the AE system. This is an obvious shortcoming of the method for applications such as OHL structures, elevator cables, and underground conductors. Based on these current technologies, there exists a need for an NDE system suitable for deployment in wire rope and metal frame structures. This paper presents the development of a diagnostic system based on electro-magnetic acoustic transducers (EMATs) integrated with informatics that would result in a smart sensor.

OPERATIONAL PRINCIPALS

The electromagnetic acoustic transducer (EMAT) couples ultrasonic energy into conductive materials. The simplest EMAT is a wire loop held near a conductive material under test with a magnet placed above the wire. The transmitter using physics similar to those acting in an electric motor develops the acoustic forces. A copper coil is placed as close to the test medium as possible and an alternating current is injected. This current produces a dynamic magnetic field **H**, which varies in time and space. The eddy current density produced in the test medium when the coil is placed near the test medium is given by Maxwell's equation:

$$J = \nabla \times H$$

From Maxwell's equations for quasi-static conditions, this eddy current flows in the medium. To create a force in the metal, permanent magnets with high magnetization are placed directly over the coil to flood the medium with magnetic flux. The eddy current interacts with the external magnetic induction B_0 to produce a force density as given below by:

$$F = J \times B_0$$

Coupled to the lattice of the metal sample, this force is called a Lorentz force and acts in a direction indicated in Figure 1. An elastic disturbance involving particle displacement **u** and velocities **du/dt** in the test specimen then propagates to the receiver.

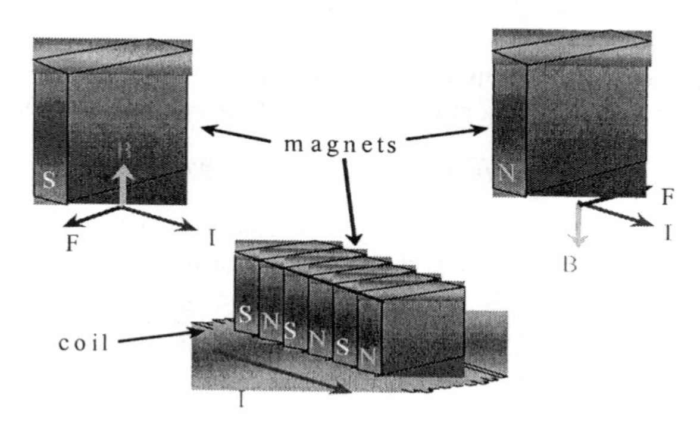

Figure 1: Lorentz Force Configuration

The receiving EMAT works similar to an electric generator. When the acoustic wave passes under the receiver, the surface of the material is displaced in the magnetic field. An electric field (E) arises according to the following equation:

$$E = \frac{du_{tot}}{dt} \times B$$

where du_{tot} / dt is the total particle velocity, which incorporates reflected, as well as incident elastic waves at the surface. With the resulting conduction current density a related field H_R and an electric field E can be generated as:

$$H_R = \left(\frac{1}{\mu_0}\right) \nabla \times A \quad \text{and} \quad E = J\,\omega\,A$$

where A is the vector electric potential outside the test specimen and around the receiver coil and μ_0 is the permeability of free space and ω is the frequency of the alternating current.

SENSOR INFORMATICS

Typical signatures received by an EMAT contain spatial, spectral and temporal information. The key for an effective utilization of this monitoring system is the signal processing and signature analysis that need to be performed on the data obtained from the EMAT receiver, as well as the self-calibration for required adjustments as the structure ages.

The Fourier transform is a powerful mathematical tool with significant physical interpretation in numerous engineering applications. However, this transformation eliminates temporal and spatial information. To recover some of that characterization, this sensory system utilizes wavelet transform as a tool for time-frequency analysis [15]. Hubbard [4] presents several criteria for choosing features of a wavelet. These include: the system of representation, regularity, vanishing moments and selectivity in frequency. To design a wavelet for transformation of the EMAT data, many of these properties have to be taken into account.

A sinusoid wavelet scaled by a normal (or Gaussian) distribution with multiple periods of the sinusoid, namely the following integral wavelet transform has been used in this research.

$$(W_\psi f)(b,a) = |a|^{-1/2} \int_{-\infty}^{\infty} f(x)\psi\left(\frac{x-b}{a}\right)dx$$

where b is the time shift and a is the dilation or scale. To capture spatial information it is proposed to incorporate a Shannon wavelet [12] of the form:

$$\psi(x) = -2\frac{\sin(2\pi x) + \cos(2\pi x)}{\pi(2x+1)}$$

IMPLEMENATION RESULTS

This section contains results of application of EMAT monitoring system to ground mat risers. Our research is continuing in other application areas for the EMAT system. Many tests are laid out to test riser configurations with cuts perpendicular to the axis of the riser, at an angle with the axis of the riser, and around the circumference of the riser. In addition, the tests are performed on types of connections. Figures 2 illustrates plots of signatures taken from riser samples with different defects. The percentage indicates the approximate amount that the cut reduces the cross-sectional area of the sample.

It can be seen from these plots that the signature changes when defects are introduced to the samples. The reflections seen in the clean sample result from the cut ends of the sample material. As a cut is introduced along the length of the sample, intermediate reflections in time are sensed by the receiver. These intermediate reflections contain information about the orientation, size, and type of defect in their spectral content.

Figure 3 contains signatures obtained from two connection configurations. The specimen with a thermoweld T-joint contains a cut in the riser material which reduces the cross sectional area by approximately 10% to 20%. This is why an intermediate reflection is seen just before the thermoweld reflection. Both of these samples also contain the flat cut ends near which the sensor is placed. This explains the reflection received on both signatures after the initial "bang." When attempting to identify defects in the material, it is important to be able to distinguish between the termination of a riser section at an appropriate connection (i.e. thermoweld joint) and a defect in the riser such as a cut in the strands. From observing the data, it appears that two measures would indicate a thermoweld joint. One such measure is the change in time width of the packet. A quantitative measure for this could be a ratio of the packet amplitude to the packet width. Another possible indicator would be a slight change in frequency. This seems plausible from the change in acoustic impedance that is represented by the thermoweld joint.

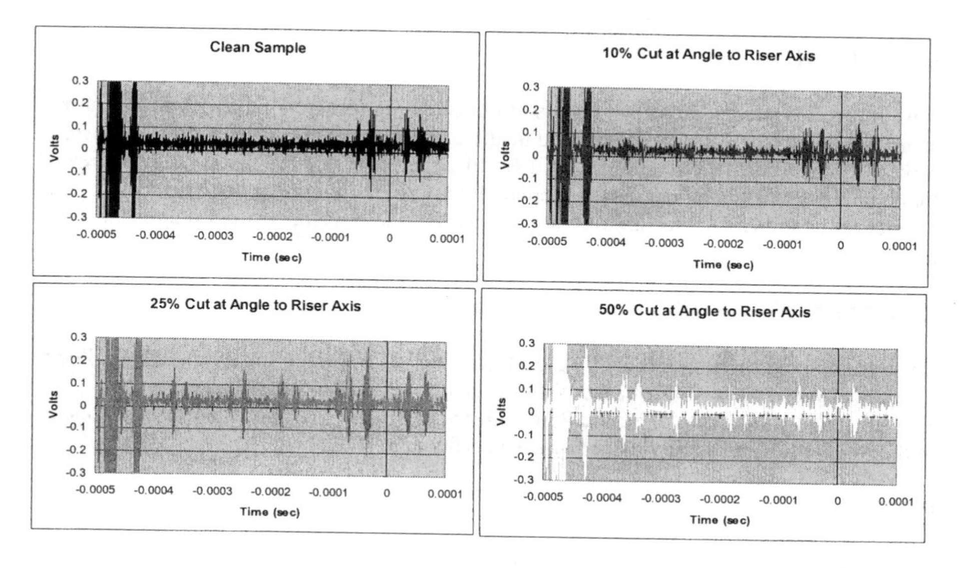

Figure 2: Data Sets for Cut at Angle to Riser Axis

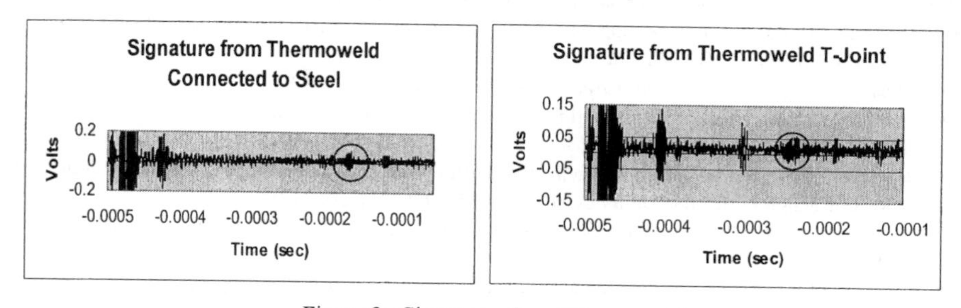

Figure 3: Signatures from Thermoweld

The characteristics of the experimental data are obviously well suited for an analysis similar to time domain reflectometry. Analyzing the data in this manner provides the time in which a reflection is received and from this the length of riser material from the receiver to the defect or connection could be calculated. With this calculation, consideration of the forward and return path should be made. To provide more information about the type of anomaly producing a reflection, spectral information is desirable. Wavelet analysis is used to provide both time and frequency based analysis. An ART-based neural network is designed as applied to this EMAT system. The combined EMAT and its informatic system, namely signal processing, wavelet transform and ART network, provide information to the service crew about the health condition of conductors, either in the ground or on a transmission line.

CONCLUSION

This paper presented some of the results of our on going research in the development of a smart sensory system for diagnostic and health assessment of metal structures and wire ropes. A new transduction mechanism that utilizes electric and magnetic fields to create torsional waves inside a metal is developed. Results from this study confirm the feasibility of integrating the EMAT technology with informatics to develop a smart diagnostic system for civil infrastructures.

ACKNOWLEDGMENT

This research has been supported by the Electric Power Research Institute (EPRI), Electric de France, Western Area Power Administration (WAPA), Tri-State Generation and Transmission, and Nebraska Public Power District (NPPD). Technical and financial support of there sponsors are acknowledged and appreciated.

REFERENCES

[1] N.F. Casey and J.L Taylor, The evaluation of wire rope by AE techniques. British Journal of NDT 27 (1985) pp 351-356.

[2] D.O. Harris and H.L. Dunegan, AE testing of wire rope, Materials Evaluation 32 (1974) pp1-6.

[3] R.E. Hobbs and M. Raoof, Interwire slippage and fatigue prediction in stranded cables for TLP tethers. *Proc. 3rd Int. Conf. Behavior of Offshore Structures* (1982) 77-92.

[4] B. Hubbard, *The World According to Wavelets.* A.K. Peters, Ltd Press. 1996.

[5] P.A.A. Laura, Acousic detectionof structural failure of mechanical cables. *J. Acoustical Society of America* **45** (1969) 791-793.

[6] P.A.A. Laure and J.R. Mathews, Monitoring the status of a mechanical cable while in operation by means of the acoustic emission method. *Ocean Engineering* **12** (1985) 211-219.

[7] P.A. Martin and J.R. Berger, Waves in wood: free vibrations of a wooden pole. Submitted for publication, May 2000.

[8] M. Raoof and I Kraincanic, Critical examination of various approaches used for analyzing helical cables. *J. Strain Analysis* **29** [9] N.L. Sao, Non-destructive evaluation of steel wire rope – Part I. *Wire Industry* September (1994) 485-488.

[10] N.L. Sao, Non-destructive evaluation of steel wire rope – Part II. *Wire Industry* January (1996) 34-3

[11] N.L. Sao, Non-destructive evaluation of steel wire rope – Part III. *Wire Industry* May (1996) 382-387.

[12] N.L. Sao, Non-destructive evaluation of steel wire rope – Part IV. *Wire Industry* November (1996) 744-799.

[13] J.L. Taylor and N.F. Casey, The acoustic emission of steel wire ropes. *Wire Industry* **51** (1984) 79-82.

[14] T.C.T. Ting, *Anisotropic Elasticity.* University Press, Oxford, 1996.

[15] H.R. Weishedel, In service inspection of wire ropes: state of the are. *Mining Science and Technology* **11** (1990) 85-93.

Seismic Protection of Historical Masonry Structures by using Control Techniques

C. A. Syrmakezis

Professor, National Technical University of Athens, Greece

ABSTRACT

A proposal for the seismic vulnerability reduction of historical masonry structures by using control techniques is described. The proposal is applied to a case study, the Minaret of Larnaca, Cyprus. Comparative analytical results for both cases, with and without vibration control devices, show the significant improvement of the structural performance of the structure, under seimic actions.

1. INTRODUCTION

The fact that a historical structure has survived in the past, does not guarantee that it will survive also in the future.

The cultural value of a historical structure, and the desire to preserve it for the future generations, demand a high level protection against any possible future destruction under future actions. Among those actions, earthquake is of primary importance.

Interventions on a historical structure is a difficult and challenging task, as several parameters have to be taken into consideration:

a) the damage sustained throughout its history, caused mainly by natural phenomena (e.g. earthquakes), may have rendered the structure incapable of withstanding future actions.

b) At the same time, any decision for interventions has to fulfil the demand not to harm what has survived till today, and not to become the cause of future damage to the structure.

To accomplish these tasks, a precise understanding of the problems to be faced by the structure, the reasons for them, as well as a sound knowledge of the effect of any intervention might have on the structure, is needed.

The assessment of the resistance of a historical structure presents special problems, not encountered in everyday structures. These problems stem from the ways in which the structures were built, the types of materials used, the complex past histories of successive changes and progressive deterioration over the years, as well as the cultural value, sometime as a work of art.

A good comprehension of the structural response, either under static or under dynamic (earthquake) actions, is a prerequisite for successful final decisions.

Within this complex outline, passive control techniques seem to be able to support the effort for the seismic protection of historical masonry structures significantly, through a (partial or even total) "over passing" of the traditional methods of interventions (through classical repair and/or strengthening techniques).

2. HISTORICAL MASONRY STRUCTURES

Masonry structure is the most common type for a historical structure.

Compared to modern structures, masonry structures present some peculiarities:

a) Masonry structures are highly vulnerable to earthquake actions

b) Stress capacity of masonry is rather low, due to its low shear and tensile strengths

c) values of masonry shear and tensile strengths present large dispersions.

d) There are uncertainties concerning type and quality of connections among the bearing elements of the structure.

e) Distribution of masses along the height of the masonry and relative insignificance of masses at the floor levels, do not allow for masonry structures assumptions reasonable for other modern structures (e.g. assumption for lumped masses at the floor levels).

3. THE ANALYTICAL PROCESS

The analysis – failure process methodology for masonry historical structures, developped last few years by the author and its group in NTUA, Greece, is used for the case under consideration in this paper (Syrmakezis et als, 1997, 1995). Namely, dynamic elastic analysis on a finite element structural model, appropriate combination for this type of structures, is used.

Design spectrum, taking into account the different soil conditions existing in the area under investigation., is supporting the dynamic spectral analysis.

For masonry structures, materials characteristics (like compressive strength, modulus of elasticity, Poisson ratio, etc), are fundamental parameters for the process application. Combination of analytical – semiempirical, and/or experimental methods are used. Tensile strength could be expressed as a percentage of the tensile strength of mortar, and therefore: $f_{wt} = \zeta \cdot f_{mt}$, where usually ζ=2/3.

The vibration control devices must be selected taking into account that masonry structures characteristics (Casciati et al. 2001, 1996, Chang. 1999). Large energy dissipation, combined with relatively small displacements, must be activated. Fluid viscous dampers (velocity depended systems) are preferred..

The response of these dampers is simulated through Maxwell model of viscoelasticity (Asano et al. 2000, Soong et al. 1990) having a non-linear damper in series with a spring as it is shown at the following figure 1.

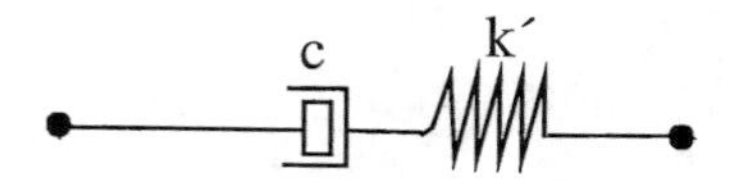

Figure 1: *Maxwell model of viscoelasticity*

The non-linear force-deformation relationship is given as following:

$$f = k' \cdot u_k = c \cdot \upsilon_c'^{\,c\exp}$$

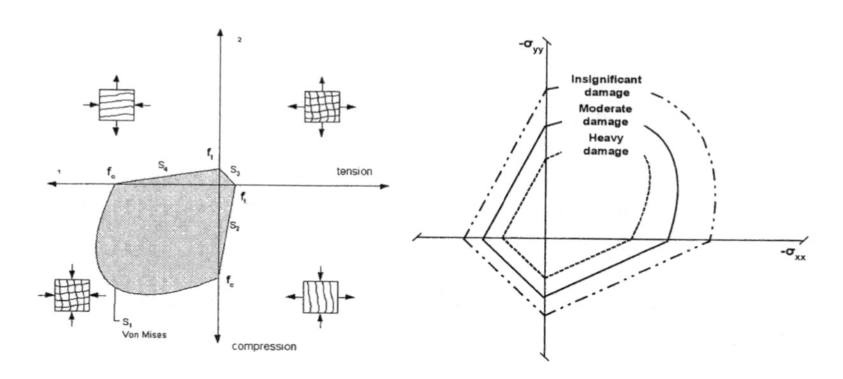

Figure 2: The modified Von Mises failure criterion - Damage levels

Fig. 3: *General view of the minaret of Larnaca, Cyprus*

The determination of failure stage of the structure due to elastic deformation is based on the Von Mises Theory (Syrmakezis et als, 2001). The modified Von Mises failure surface is formed by the interaction of four surfaces S1, S2, S3 and S4 as presented at figure 2 (level σ_{xx}, σ_{yy}).

4. THE CASE OF THE LARNACA, CYPRUS MINARET

The process described in the previous paragraph, is applied to the case of a real masontry structure with specific characteristics and demands: A minaret, part of an islamic mosque, located in the city of Larnaca, Cyprus (fig. 3).

The structural system consists of an external cylindrical shell (large stones and thin joints of mortar) and internal stone stairs connected to the external shell. The total height of the structure is 26m.

The internal diameter from the base level to the ¾ of the total high the structure (at about 19.5m, where the balcony exists), is aproximately 2m. For the rest (higher part) of the structure the diameter is about 1.8m. The thickness of the external shell is variable:

65cm from the base to the entrance level (1/4 of total height, about 6.5m),
55cm from entrance level to the balcony level (from 6.5m to 19.5m levels). and,

Fig. 4: *The middle part of the minaret, with vertical cracks on the masonry.*

Fig. 5: *The upper part of the minaret, with vertical cracks on the masonry.*

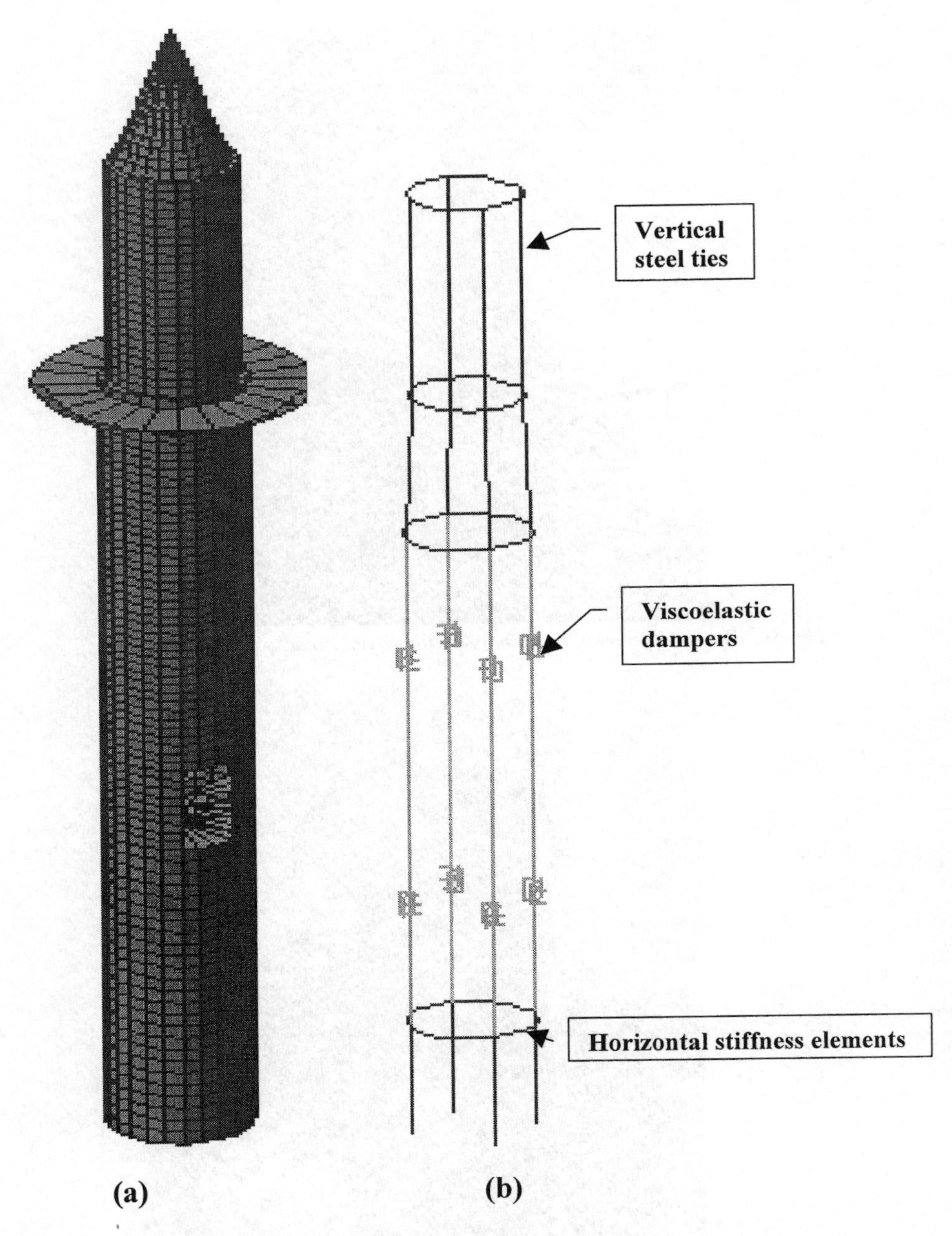

Fig. 6. *General layout of the finite element model (a). Positions of the vicoelastic dampers (b).*

45cm from the balcony level to the top (from 19.5m to 26m levels).

The ratio of height to base is greater than 5 (h/b>5) and therefore the structure is very flexible.

The lower part of the minaret, from ground level to entrance level (+6.5m), is more rigid than the rest of the structure and is formed by an irregular structural system, filled with mixed stones and lime or gypsum mortar.

The area of Larnaka is an earthquake prone area. A recent strong earthquake event on 9/10/1996, gave a magnitude of Ms = 6.3.

The design spectrum of Cyprus Aseismic Code with the consistent accelerograms were taken into account. The following figure presents the seismic zones of Cyprus and the shape of the spectrum. According to these data, for seismic zone 5 (Larnaca area) and pick ground acceleration equal to 0.15g, a design acceleration of 0.27g is taken into consideration.

The Finite Element Analysis model of the structure was developed at three dimensions and consists of 2984 shell elements, 3097 joints and 82 (+ 8 for the dampers) bar elements. For a reliable determination of the system deformation at three dimensions, six degrees of freedom were considered, with three translations and three rotations.The general layout of the mesh is presented in fig. 6. The finite element mesh was developed in such a way that the ideal massing at the joints of the structure would result to a more realistic modeling of the inertia forces. The vibration control devices used here are fluid viscous dampers. Dampers have been inserted for the analysis externaly, in the vertical direction, across the height of the structure, in pairs and in four columns (fig. 6b).

Two steps for the analysis have been followed: one without the dampers and one with the dampers applied on the structure. The reduction of the number of points collapsed is significant for the second case. Remarcable is also for the second case the diminuation in both directions of the horizontal displacement, at the top of the structure (about 30%). Comparative results are given on the table below.

	Ux	Uy
Existing structure	85 mm	98 mm
structure with control devices	60 mm	62 mm
reduction	29,4%	36,7%

Although several problems (mainly technical) have to be solved prior to an application of the method on a real structure, first results are promising.

5. CONCLUSIONS

The use of vibration control devices can significantly reduce seismic vulnerability, leading to an alternative method of strengthening historical structures against dynamic (earthquake) loads.

This use of vibration control devices will be imperative, when reversibility of interventions will be required.

REFERENCES

Asano, K. and Nakagawa, H. (2000) Optimum Seismic Response Control of Structural System Using Maxwell-Type Nonlinear Viscous Damper Based on Random Earthquake Response. *Proceedings of the 3rd International Workshop on Structural Control, Paris, France, pp 39-48.*

Casciati, F. and Faravelli, L. (2001) Stochastic Nonlinear Controllers, *Proceedings IUTAM Symposium on Non Linearity and Stochastic Structural Dynamics, Madras, Kluwer.*

Casciati, F. and Lagorio, H.J. (1996) Urban Renewal Aspects and Technological Devices in Infrastructure Rehabilitation. *Proceedings of the First European Conference on Structural Control, Barcelona, pp 173-181.*

Soong, T.T. and Dargush, G.F. (1990) Passive Energy Dissipation Systems in Structural Engineering. *John Wiley & Sons, New York, USA.*

Syrmakezis C., Sophocleous A.,Kordouli V. (2001) Seismic vulnerability reduction of historical masonry structures using vibration control devices. *Proceedings of the European meeting on intelligent structures,EMIS 2001, ischia, Italy.*

Syrmakezis C., Asteris P., Sophocleous A. (1997) Earthquake Resistant Design of Masonry Tower Structures. *Proceedings, 5th STREMA, St. Sebastian.*

Syrmakezis C., Sophocleous A., Asteris P., Liolios A. (1996) Earthquake Resistant Design of Masonry Structural System. *Proceedings of the Earthquake Resistant Engineering Structures, ERES96, Thessaloniki, Greece.*

Syrmakezis C., Chronopoulos M., Sophocleous A., Asteris P (1995) Structural Analysis Methodology for Historical Buildings. *Proceedings, 4th STREMA, Creta.*

Passivity based Control of Piezoelectric Structures

Andreas Kugi* and Kurt Schlacher†

Abstract

This contribution deals with the application of passivity based control to composite piezoelectric structures. The constructional design of the layer structure, the form of the electrodes or the choice of the polarization offer an additional degree of freedom for the control loop design. Thus, it will be shown that an appropriate voltage supply of the actuator layers in combination with suitable sensor layers provide a collocated actuator-sensor configuration. This serves as a basis for a simple passivity based control law with output feedback. It can be proven that for the bending of a piezoelectric composite cantilever beam this controller design together with a special layer structure leads to an exponentially stable closed loop system.

1 Introduction

Passivity based control is well established in the field of finite dimensional dynamic systems. The main advantage of this approach is that the design of the controller is directly related to a proof of the stability of the closed loop, see, e.g., (7), (14), (18), (21). The concept of the so called actuator/sensor collocation, which constitutes an energy port of the system, is of great importance, since this principle can be extended from finite to infinite dimensional systems and therefore also to piezoelectric composite structures. Additionally, the collocated actuator/sensor configuration together with a suitable controller design prevent the so called "spillover"-effects of the closed loop (4).

Many ideas of passivity based control can be generalized from the finite to the infinite dimensional case in a straightforward manner, but stability investigations become much more sophisticated. According to the knowledge of the authors, there exists no general stability theory for arbitrary distributed parameter systems like it exists for the lumped parameter case. The interested reader may, e.g., consult (9), (2) and the literature cited therein for more information on this topic.

Piezoelectric composite structures, like beams and plates, are certainly members of the class of infinite dimensional systems. Since these structures offer the possibility to choose the spatial distribution of the electrodes, one gains a new degree of freedom for the design of the structure, see, e.g., (3), and for the design of the control law, e.g., (7), (16), (20). One of the main goals of this contribution is to show by means

*Prof. A. Kugi is head of the Chair of System Theory and Automatic Control at the University of Saarland, Germany.

†Prof. K. Schlacher is head of the Department of Automatic Control and Control Systems Technology and of the Christian Doppler Laboratory for Automatic Control of Mechatronic Systems in Steel Industries, both located at the Johannes Kepler University of Linz, Austria.

of a piezoelectric composite beam that the problem of the actuator/sensor collocation can be solved independently of the boundary conditions, provided that special active layer pairs can be combined with suitable power supplies. Within the linear theory, we can characterize a piezoelectric composite beam which allows us to influence the longitudinal and the vertical vibrations in a desired fashion. Furthermore, it can be shown that all controllers designed on the basis of the collocation method for a special cantilever beam lead to a closed loop which is exponentially stable.

The contribution is organized as follows: Section 2 summarizes some results of passivity based control and the concept of collocation for the finite-dimensional and the extension to the infinite-dimensional case. Section 3 is dedicated to piezoelectric composite structures. The equations of motion for a piezoelectric composite Euler-Bernoulli beam are derived based on the constitutive relations of a linear piezoelectric material. Additionally, the influence of the spatial distribution of the electrodes on the properties of the actuators and sensors is studied. Section 4 shows by means of a symmetric multi-layered beam that the actuators and sensors can be designed in such a way that they provide a collocated configuration independent of the boundary conditions. Section 5 deals with the transversal vibrations of a piezoelectric composite cantilever beam, where the collocation of the actuator and the sensor is chosen to achieve exponential stability with a simple feedback law. The main results are shortly summarized in Section 6.

2 Passivity based control

We will shortly discuss the class of the so called finite-dimensional **PCH** (**P**ort **C**ontrolled **H**amiltonian) systems, see, e.g., (10), (21). To start with, we consider the Hamiltonian system

$$q_1^\alpha = \delta^{\alpha\beta}\bar{\partial}_\beta H \; , \quad p_1^\alpha = -\delta^{\alpha\beta}\partial_\beta H \; , \quad \partial_\alpha = \frac{\partial}{\partial q^\alpha} \; , \quad \bar{\partial}_\beta = \frac{\partial}{\partial p^\beta} \tag{1}$$

with the finite dimensional state (q^α, p^α), $\alpha, \beta = 1, \ldots, n$ in canonical coordinates with the Hamiltonian function $H = H(q, p)$ and the Kronecker symbol $\delta^{\alpha\beta}$. The generalized positions are denoted by q and the generalized momenta by p. Let us introduce the vector field or the differential operator[1] d_0,

$$\mathrm{d}_0 = \partial_0 + q_1^\alpha \partial_\alpha + p_1^\alpha \bar{\partial}_\alpha \; , \quad \partial_0 = \frac{\partial}{\partial t} \; . \tag{2}$$

Like in the formula above, we will use Einstein's convention for sums to keep the formulas short and readable, whenever the range of the indices is clear. The functions $(\bar{q}(t), \bar{p}(t))$ are called trajectories of (1), iff $q^\alpha = \bar{q}^\alpha(t)$, $p^\alpha = \bar{p}^\alpha(t)$, $q_1^\alpha = \partial_0 \bar{q}^\alpha(t)$, $p_1^\alpha = \partial_0 \bar{p}^\alpha(t)$ meet (1). Obviously, the Hamiltonian H meets $\mathrm{d}_0 H = 0$, iff (1) is met, or H is constant along the trajectories of (1). In the case of a PCH-system with inputs e_η, $\eta = 1, \ldots, m$, the corresponding Hamiltonian function H takes the form

$$H = H_0 - H_\eta e^\eta \; , \tag{3}$$

[1]The advantage of the notation d_0, q_1^α, p_1^β will be clarified, if we turn to infinite-dimensional systems.

with the *Hamiltonian of the free system* H_0 and the so called *interaction Hamiltonians* H_η (11). On a trajectory, where (1) is met, we get

$$\mathrm{d}_0 H_0 = \partial_0 H + (\mathrm{d}_0 H_\eta)\, e^\eta + H_\eta \partial_0 e^\eta = -H_\eta \partial_0 e^\eta + (\mathrm{d}_0 H_\eta)\, e^\eta + H_\eta \partial_0 e^\eta = (\mathrm{d}_0 H_\eta)\, e^\eta \quad (4)$$

for $e^\eta = e^\eta(t)$. If H_0 is the stored energy, then (4) states that the change of H_0 equals the power supplied by the input e^η. The choice of the output y_η given by

$$y_\eta = H_\eta\ , \quad \eta = 1, \ldots, m \quad (5)$$

is called the *output collocated* with the input e^η. Equivalently, the expression $\dot{y}_\eta e^\eta$,

$$\dot{y}_\eta = \mathrm{d}_0 H_\eta\ , \quad (6)$$

together with (1) is the power flow across the system borders. In the language of network theory the pairs $(e^\eta, \dot{y}_\eta)$ are called ports, see, e.g., (6).

We assume without any restriction of generality that $(q,p) = (0,0)$ is an equilibrium of the free system. Let us choose the simple feedback law

$$e^\eta = -K^{\eta\zeta} \dot{y}_\zeta\ , \quad \zeta = 1, \ldots, m \quad (7)$$

with the positive semi-definite matrix $\left[K^{\eta\zeta}\right]$, then the relation (4) implies

$$\mathrm{d}_0 H_0 = -\dot{y}_\zeta K^{\zeta\eta} \dot{y}_\eta \le 0\ . \quad (8)$$

In the finite-dimensional case the positive definiteness of H_0 implies the stability of the equilibrium in the sense of Lyapunov. Additionally, one can apply the invariance principle of Karassowskij-LaSalle to check for the asymptotic stability, see, e.g., (5). This approach can be extended to dynamic feedback laws in the following manner, see, Fig. 1. Let the linear controller be given by

$$\begin{aligned} z_1^{\alpha_c} &= A^{\alpha_c}_{\beta_c} z^{\beta_c} + B^{\alpha_c \zeta} \dot{y}_\zeta\ , \quad \alpha_c, \beta_c = 1, \ldots, n_c \\ -e^\eta &= C^\eta_{\alpha_c} z^{\alpha_c} + D^{\eta\zeta} \dot{y}_\zeta \end{aligned} \quad (9)$$

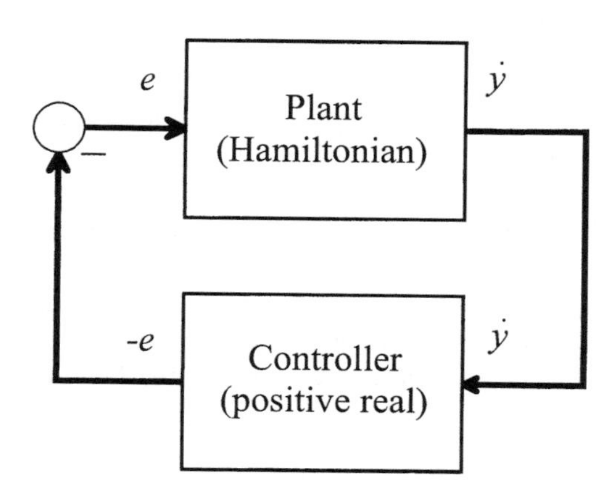

Figure 1: Feedback Interconnection of a PCH-System and a Positive Real Controller.

with the matrices $A = \left[A^{\alpha_c}_{\beta_c}\right]$, $B = \left[B^{\alpha_c \zeta}\right]$, $C = \left[C^{\eta}_{\alpha_c}\right]$ and $D = \left[D^{\eta\zeta}\right]$ of real numbers. If the controller transfer matrix $\left[R^{\eta}_{\zeta}(s)\right]$ of (9) is positive real, then we can find a positive definite matrix $P = [P_{\alpha_c \beta_c}]$ and matrices L, W such that the equations

$$\begin{aligned} PA + A^T P &= -L^T L \\ PB &= C^T - L^T W \\ W^T W &= D^T + D \end{aligned} \tag{10}$$

are met according to the Kalman-Yakubovich-Popov lemma (5). Using V,

$$V(q, p, z) = H_0 + \frac{1}{2} z^{\alpha_c} P_{\alpha_c \beta_c} z^{\beta_c} \tag{11}$$

as a candidate for a Lyapunov function of the closed loop, we get

$$\begin{aligned} \mathrm{d}_0 V &= \dot{y}_\eta e^\eta + \tfrac{1}{2} z_1^{\alpha_c} \left(P_{\alpha_c \beta_c} + P_{\beta_c \alpha_c}\right) z^{\beta_c} \\ &= -\dot{y}^T \left(Cz + D\dot{y}\right) + \tfrac{1}{2}\left(z^T P \left(Az + B\dot{y}\right) + \left(z^T A^T + \dot{y}^T B^T\right) Pz\right) \\ &= -\tfrac{1}{2}\dot{y}^T \left(D + D^T\right)\dot{y} + \tfrac{1}{2}\dot{y}^T\left(B^T P - C\right) z + \tfrac{1}{2} z^T \left(PB - C^T\right)\dot{y} \\ &\quad + \tfrac{1}{2} z^T \left(PA + A^T P\right) z \\ &= -\tfrac{1}{2}\left(z^T L^T + \dot{y}^T W^T\right)\left(Lz + W\dot{y}\right) \le 0 \end{aligned} \tag{12}$$

with

$$\mathrm{d}_0 = \partial_t + q_1^\alpha \partial_\alpha + p_1^\alpha \bar{\partial}_\alpha + z_1^{\alpha_c} \partial_{\alpha_c}\,, \qquad \partial_{\alpha_c} = \frac{\partial}{\partial z^{\alpha_c}} \tag{13}$$

where (1) and (9) are met. Thus, the equilibrium of the closed loop is stable for any positive definite function H_0.

Let us extend our approach from the finite-dimensional case to infinite-dimensional systems. Now, the independent variables are the time $t = x^0$ and the spatial variables x^i, $i = 1, \ldots, r$. We assume that (x) are local coordinates of an r-dimensional manifold $\mathcal{D}$ and we form the base manifold $\mathcal{B} = R \times \mathcal{D}$ with the local variables (t, x), $t \in \mathbb{R}$. The dependent variables are denoted by w^α, $\alpha = 1, \ldots, s$, and we introduce the $(1 + r + s)$-dimensional total manifold $\mathcal{E}$ with the local coordinates (t, x, w) and the projection $\pi : \mathcal{E} \to \mathcal{B}$ locally given by $\pi(t, x, w) = (t, x)$. A local section σ of $\mathcal{E}$ is a map $\sigma : \mathcal{B} \to \mathcal{E}$, which meets $\pi \circ \sigma(t, x) = (t, x)$. Let σ be a smooth section, then we can determine all possible partial derivatives given by

$$(\partial_0)^{j_0} \cdots (\partial_r)^{j_r} \sigma^\alpha = \sigma^\alpha_J\,, \quad \sigma_{(0,\cdots,0)} = \sigma\,, \quad \partial_i = \frac{\partial}{\partial x^i} \tag{14}$$

with $x^0 = t$. The index $J = (j_0, \ldots, j_r)$ is an ordered multi-index of the order $\#J = \sum_{k=0}^{r} j_k$. Without discussing the details, we refer the reader to the literature, e.g., (13), (16). The manifold $J^n(\mathcal{E})$ with local coordinates $\left(t, x, w^{(n)}\right)$, where $w^{(n)}$ is a shortcut for (w^α_J), $\alpha = 1, \ldots, s$, $\#J = 0, \ldots, n$ is called the n-th *Jet-Manifold* $J^n(\mathcal{E})$ of $\mathcal{E}$. Let σ be a section of $\mathcal{E}$, then $w^\alpha_J = \sigma^\alpha_J$ is obviously a section of $J^n(\mathcal{E})$ denoted by $j^{(n)}(\sigma)$. We can define the differential operators d_k,

$$\mathrm{d}_k = \partial_k + w^\alpha_{J+1_k} \partial^J_\alpha\,, \quad \partial^J_\alpha = \frac{\partial}{\partial w^\alpha_J}\,, \quad \#J = 0, \ldots, n\,, \quad k = 0, \ldots, r \tag{15}$$

on $J^{n+1}(\mathcal{E})$, where 1_k denotes the index (δ_{kl}), $l = 0, \ldots, r$, and $J + 1_k$ the index $(j_l + \delta_{kl})$. The operators d_k are also called the total derivatives with respect to the independent variables x^k. Obviously, the operators d_0 of (2), (13) are nothing else than the total derivatives with respect to t. The *Euler-Lagrange* operator is defined by

$$\mathsf{E}_\alpha = (-1)^{\#J} \mathrm{d}_J \partial_\alpha^J \ , \quad \mathrm{d}_J = \mathrm{d}_{j_0} \cdots \mathrm{d}_{j_r} \ , \tag{16}$$

where the sum is taken over all possible multi-indices $1 \leq \#J \leq n$.

Subsequently, we confine ourselves to the following special case. The number of inputs e^η, $\eta = 1, \ldots, m$, is finite and the Hamiltonian is given by (3), but the functions H_0, H_η are replaced by the functionals H_0, H_η,

$$H_0 = \int_{\mathcal{D}} h_0 \left(x, w^{(n)}\right) \mathrm{d}v \ , \quad H_\eta = \int_{\mathcal{D}} h_\eta \left(x, w^{(n)}\right) \mathrm{d}v \tag{17}$$

with $\mathrm{d}v = \mathrm{d}x^1 \wedge \cdots \wedge \mathrm{d}x^r$ and suitable boundary conditions for $w^{(n)}$ on the boundary $\partial\mathcal{D}$ of $\mathcal{D}$, see, e.g., (13), (17). Let us assume that we can find canonical coordinates $(q, p) = (w)$, $\alpha = 1, \ldots, s/2$, $h_0 = h_0\left(x, q^{(n)}, p\right)$, $h_\eta = h_\eta\left(x, q^{(n)}, p\right)$, then the equations of motion read as

$$q_{1_0}^\alpha = \delta^{\alpha\beta} \bar{\mathsf{E}}_\beta \left(h_0 - h_\eta e^\eta\right) \ , \quad p_{1_0}^\alpha = -\delta^{\alpha\beta} \mathsf{E}_\beta \left(h_0 - h_\eta e^\eta\right) \ , \tag{18}$$

where E_β, $\bar{\mathsf{E}}_\beta$ denote the Euler-Lagrange operators with respect to q^α and p^α. It is worth mentioning that canonical coordinates may not exist in contrast to the finite-dimensional case (13). Analogously to (5) we introduce the output

$$y_k = H_\eta = \int_{\mathcal{D}} h_\eta \mathrm{d}v \tag{19}$$

collocated to the input e^η, which meets

$$\mathrm{d}_0 H_0 = \mathrm{d}_0 \left(H_\eta\right) e^\eta \tag{20}$$

with d_0 from (15) combined with (18). Obviously, the relation (17) corresponds to the equation (4) for the finite-dimensional case.

Unfortunately, the results concerning the stability cannot be extended to the infinite-dimensional case in a straightforward manner. Even if the Hamiltonian functional of the free system H_0 is positive definite and the calculated controller of the form (7) guarantees $\mathrm{d}_0 H_0 \leq 0$, this provides only a necessary csondition for the stability of the closed loop (1), (9), in contrast to the finite-dimensional case. Until now, the authors do not know, except for a huge amount of more or less restrictive special cases, a generally applicable stability test.

3 The Mathematical Model

Since this contribution deals mainly with control problems of smart structures, we confine the mechanical model to a thin linear piezoelectric multi-layered beam, where the Euler-Bernoulli hypothesis is fulfilled. From now on, $\{e_1, e_2, e_3\}$ denotes the canonical orthogonal basis of the 3-dimensional Euclidian space with the coordinates x^i, $i = 1, 2, 3$, which meets $(e_i, e_j) = \delta_{ij}$ for the standard inner product $(\cdot, \cdot)$.

3.1 Linear Piezoelectricity

The linear constitutive equations of the piezoelectric material are given by

$$\sigma^{ij} = c^{ijkl}\varepsilon_{kl} - a_k^{ij} D^k \tag{21a}$$

$$E_i = -a_i^{kl}\varepsilon_{kl} + d_{ik}D^k \tag{21b}$$

with the mechanical stress $\sigma = \sigma^{ij}\partial_i \otimes \partial_j$, the strain $\varepsilon = \varepsilon_{ij}\mathrm{d}x^i \otimes \mathrm{d}x^j$, the electric field $E = E_i\mathrm{d}x^i$ and the electric flux density $D = D^i\partial_i \rfloor \mathrm{d}v$, $\mathrm{d}v = \mathrm{d}x^1 \wedge \mathrm{d}x^2 \wedge \mathrm{d}x^3$ (12). Here, the symbols $\otimes$ and $\rfloor$ denote the tensor product and the interior product. The integrability conditions

$$c^{ijkl} = c^{jikl} = c^{ijlk} = c^{klij} \ , \quad a_j^{ki} = a_i^{kj}, \quad d_{ij} = d_{ji} \tag{22}$$

ensure the existence of the energy function W_p

$$W_p = \int_{\mathcal{V}} \left(\frac{1}{2}\varepsilon_{ij}c^{ijkl}\varepsilon_{kl} - \varepsilon_{ij}a_k^{ij} D^k + \frac{1}{2}D^i d_{ik} D^k \right) \mathrm{d}v \ , \tag{23}$$

where $\mathcal{V}$ denotes the volume of the smart structure. Since the piezoelectric material is free of charge, the electric flux density D meets the relation

$$\partial_i D^i = 0 \ . \tag{24}$$

Further assumptions concerning the equations (21) and (23) follow from the beam or plate structure under investigation.

3.2 The Piezoelectric Euler-Bernoulli Beam

We consider the piezoelectric multi-layered beam of Fig. 2 and denote the displacement of a point $x^i e_i$ by (u^i). According to the Euler-Bernoulli hypothesis we set

$$\varepsilon_{11} = \bar{u}_{01}^1 - x^3 \bar{u}_{02}^3 \ , \tag{25}$$

where $(\bar{u}^i)$, $\bar{u}^2 = 0$, is the displacement of the reference line. The independent coordinates are (t, x^1), see, Fig. 2. Furthermore, we assume

$$\varepsilon_{kl} = 0 \ , \quad k + l > 2 \ , \qquad D^1 = D^2 = 0$$

and derive the simplified constitutive equations, see (21),

$$\sigma^{11} = c^{1111}\varepsilon_{11} - a_3^{11} D^3 \tag{26a}$$

$$E_3 = -a_3^{11}\varepsilon_{11} + d_{33}D^3 \ . \tag{26b}$$

The material parameters c^{1111}, a_{11}^3, d_{33} may vary from layer to layer but they are supposed to be constant within a single layer. Combining the relations (25) and (26) with (23), we derive the stored energy of the beam W_p,

$$W_p = \frac{1}{2}\int_{\mathcal{V}} \left(\overbrace{c^{1111}\left(\bar{u}_{01}^1 - x^3\bar{u}_{02}^3\right)^2}^{A} - \overbrace{2a_3^{11}\left(\bar{u}_{01}^1 - x^3\bar{u}_{02}^3\right) D^3}^{B} + \overbrace{d_{33}\left(D^3\right)^2}^{C} \right) \mathrm{d}v \ . \tag{27}$$

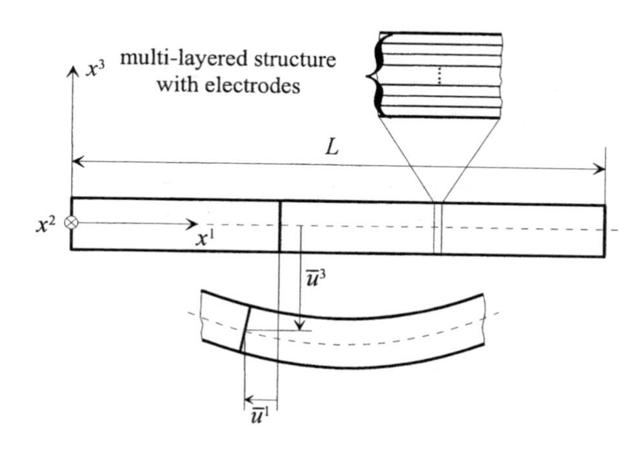

Figure 2: Piezoelectric Multi-Layered Beam.

The expression A represents the energy density of the mechanical part, the expression B stands for the coupling terms and C for the electrical part. Neglecting the rotational inertia, we derive the kinetic energy W_k as

$$W_k = \frac{1}{2\rho}\int_{\mathcal{V}}\left(\left(p^1\right)^2 + \left(p^3\right)^2\right)\mathrm{d}v\ , \qquad p^1 = \rho u^1_{10}\ , \quad p^3 = \rho u^3_{10} \tag{28}$$

with the mass density ρ. Since the beam consists of active and substrate layers, we must set $D^3 = 0$ for the electrically inactive part.

3.3 Piezoelectric Actuator Layers

The electrodes of an actuator layer are connected to a power supply with the voltage U^η. Using the relation (26b) and neglecting $a_3^{11}\varepsilon_{11}$ in comparison to $d_{33}D^3$, see (19), we get

$$U^\eta = \int_{h_\eta}^{h_{\eta+1}} E_3\mathrm{d}x^3 = d_{33}\left(h_{\eta+1} - h_\eta\right)D^3$$

according to Fig. 3. Let us assume that the layers of the beam are built up symmetrically with respect to the $\left(x^1, x^2\right)$-plane. We can choose a symmetric or a skew symmetric power supply for a pair of two symmetric layers, see Fig. 3. In this case, one can cancel out several terms of the coupling energy, see (27). Another possibility to achieve special actuator properties is given by a certain spatial distribution $B\left(x^1\right)$ of the metallic electrodes, as shown in Fig. 4.

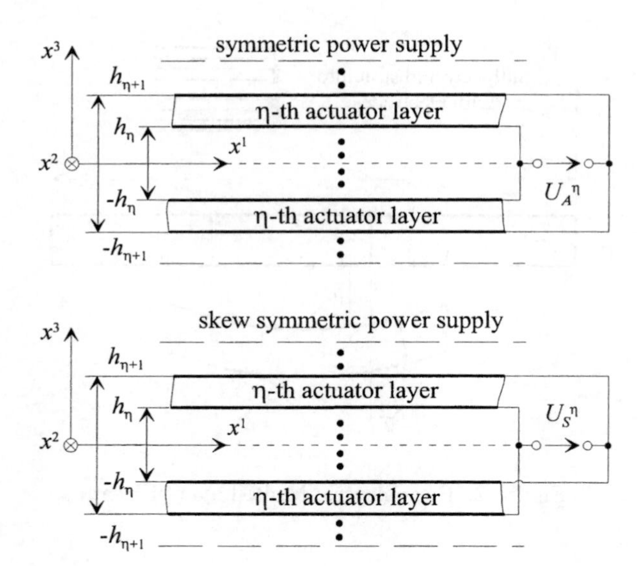

Figure 3: Symmetric and Skew Symmetric Power Supply.

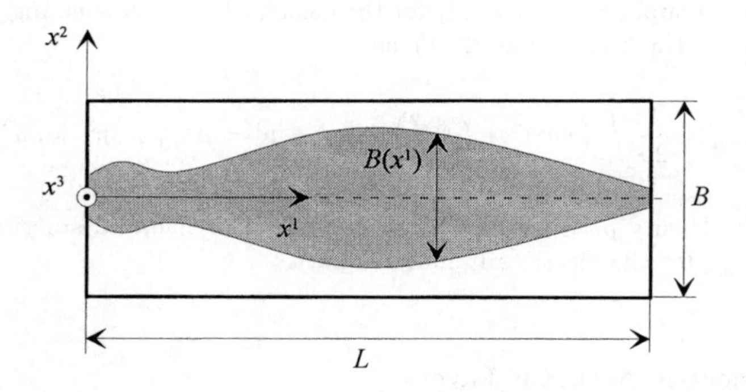

Figure 4: Spatial Distribution of the Surface of the Electrodes.

3.4 Piezoelectric Sensor Layers

The sensor layers under investigation are assumed to be short circuited, which implies

$$\int_{h_\eta}^{h_{\eta+1}} E_3 \mathrm{d}x^3 = 0 \ . \tag{29}$$

Because of the small layer thickness $(h_{\eta+1} - h_\eta)$ of a piezoelectric sensor layer, we assume that E_3 is constant, which implies $E_3 = 0$ because of (29). Therefore, we may treat the sensor layers like substrate layers for the determination of the stored energy (27).

3.5 A Special Stored Energy Function

We assume that the beam consists of $2m$ symmetric layers. Furthermore, $2m_S$ actuator layers with the width $B_\eta^S(x^1)$ have a symmetric power supply with the voltage U_S^η, $\eta = 1,\ldots,m_S$, and $2m_A$ actuator layers with the width $B_\eta^A(x^1)$ have a skew symmetric power supply with the voltage U_A^η, $\eta = 1,\ldots,m_A$. In this case the stored energy of the beam W_p, see (27), simplifies to

$$W_p = \frac{1}{2}\int_0^L \Lambda_1 \left(\bar{u}_{01}^1\right)^2 + \Lambda_2 \left(\bar{u}_{02}^2\right)^2 \mathrm{d}x_1 - U_S^\eta \int_0^L \Lambda_\eta^S\left(x^1\right)\bar{u}_{01}^1 \mathrm{d}x_1 + U_A^\eta \int_0^L \Lambda_\eta^A(x^1)\bar{u}_{02}^3 \mathrm{d}x_1, \tag{30}$$

with

$$\Lambda_1 = \sum_{j=1}^{2m}\int_{A_j} c^{1111}\mathrm{d}x^2 \wedge \mathrm{d}x^3 \ , \quad \Lambda_2 = \sum_{j=1}^{2m}\int_{A_j} c^{1111}\left(x^3\right)^2 \mathrm{d}x^2 \wedge \mathrm{d}x^3 \tag{31}$$

as well as

$$\Lambda_\eta^S\left(x^1\right) = \frac{2a_3^{11}}{d_{33}} B_\eta^S\left(x^1\right) \ , \quad \Lambda_\eta^A(x^1) = \frac{a_{311}\left(h_{\eta+1} + h_\eta\right)}{d_{33}} B_\eta^A\left(x^1\right) \ , \tag{32}$$

where A_j denotes the cross section of the j-th layer. All other indices of the material parameters are suppressed. Finally, the kinetic energy W_k, see (28), follows as

$$W_k = \frac{1}{2\mu}\int_0^L \left(\left(\bar{p}^1\right)^2 + \left(\bar{p}^3\right)^2\right)\mathrm{d}x^1 \ , \quad \bar{p}^1 = \mu\bar{u}_{10}^1 \ , \quad \bar{p}^3 = \mu\bar{u}_{10}^3, \quad \mu = \sum_{j=1}^{2m} A_j\rho_j \ . \tag{33}$$

3.6 The Equations of Motion

The equations of motion of the beam can be derived in a straightforward manner by Hamilton's principle from (30), (33) in the following form

$$\mu\bar{u}_{20}^3 + \Lambda_2\bar{u}_{04}^3 + (\partial_1)^2\Lambda_\eta^A(x^1)U_A^\eta = 0 \ , \quad \mu_{20}^1 - \Lambda_1\bar{u}_{02}^1 + \partial_1\Lambda_\eta^S(x^1)U_S^\eta = 0 \ . \tag{34}$$

The way how the beam is supported at the two edges $x^1 = 0$ and $x^1 = L$ determines the kinematic boundary conditions. Combining the possible kinematic boundary conditions with the dynamic ones, we derive the following generalized relations

$$\left(\Lambda_1\bar{u}_{01}^1 - \Lambda_\eta^S\left(x^1\right)U_S^\eta\right)\bar{u}^1 = 0 \tag{35a}$$

$$\left(\Lambda_2\bar{u}_{02}^3 + \Lambda_\eta^A(x^1)U_A^\eta\right)\bar{u}_{01}^3 = 0 \tag{35b}$$

$$\left(\Lambda_2\bar{u}_{03}^3 + \partial_1\Lambda_\eta^A(x^1)U_A^\eta\right)\bar{u}^3 = 0 \ , \tag{35c}$$

which must be met at $x^1 \in \{0, L\}$, where L denotes the length of the beam. Additionally, the relations (34) and (35a) show that one can affect the longitudinal vibrations by U_S^η and the vertical ones by U_A^η.

4 Actuator/Sensor Collocation

According to the considerations above, the piezoelectric composite beam is an infinite-dimensional PCH-system with the Hamiltonian H,

$$H = H_0 - \sum_{\eta=1}^{m_S} H_\eta^S U_S^\eta - \sum_{\eta=1}^{m_A} H_\eta^A U_A^\eta \tag{36}$$

and

$$H_0 = \frac{1}{2}\int_0^L \left(\Lambda_1 \left(\bar{u}_{01}^1\right)^2 + \Lambda_2 \left(\bar{u}_{02}^3\right)^2\right) \mathrm{d}x^1 + \frac{1}{2\mu}\int_0^L \left(\left(\bar{p}^1\right)^2 + \left(\bar{p}^3\right)^2\right) \mathrm{d}x^1 \tag{37}$$

$$H_\eta^S = \int_0^L \Lambda_\eta^S \left(x^1\right) \bar{u}_{01}^1 \mathrm{d}x_1 \,, \qquad H_\eta^A = -\int_0^L \Lambda_\eta^A(x^1)\bar{u}_{02}^3 \mathrm{d}x_1 \,. \tag{38}$$

The equations of motion (34) in canonical coordinates $q^1 = \bar{u}^1$, $q^2 = \bar{u}^3$, $p^1 = \bar{p}^1$, $p^2 = \bar{p}^3$ in Hamiltonian form follow directly from (36) together with (18). The change of the energy of the free system H_0 along a trajectory is given by

$$\begin{aligned} \mathrm{d}_0 H_0 \quad = \quad & -U_A^\eta \textstyle\int_0^L \left(\partial_1\right)^2 \Lambda_\eta^A(x^1)\bar{u}_{10}^3 \mathrm{d}x^1 + U_A^\eta \left\{\partial_1 \Lambda_\eta^A(x^1)\bar{u}_{10}^3 - \Lambda_\eta^A(x_1)\bar{u}_{11}^3\right\}_{x^1\in\{0,L\}} \\ & -U_S^\eta \textstyle\int_0^L \partial_1 \Lambda_\eta^S \left(x^1\right) \bar{u}_{10}^1 \mathrm{d}x_1 + U_S^\eta \left\{\Lambda_\eta^S \left(x^1\right) \bar{u}_{10}^1\right\}_{x^1\in\{0,L\}} \,, \end{aligned} \tag{39}$$

where we used the relations (35) and the integration by part technique. Obviously, the relation (39) shows how the flow of the power can be modified by the actuator voltages U_A^η and U_S^η and by a suitable choice of the widths of the electrodes for $\Lambda_\eta^A(x^1)$ and $\Lambda_\eta^S(x^1)$, see (32).

If the sensor configuration is chosen according to the Subsection 3.4, then we derive the relations for the electrical charge Q_η,

$$Q_\eta = \int_{\mathcal{A}} D^3 \mathrm{d}x^1 \wedge \mathrm{d}x^2 = \int_0^L \int_{B(x^1)} \frac{a_3^{11}}{d_{33}} \varepsilon_{11} \mathrm{d}x^1 \wedge \mathrm{d}x^2 = \int_0^L \frac{a_3^{11}}{d_{33}} B\left(x^1\right)\left(\bar{u}_{01}^1 - \bar{h}\bar{u}_{02}^2\right) \mathrm{d}x_1 \tag{40}$$

from (26b), (25) by integration over the effective surface $\mathcal{A}$ of the electrodes with $\bar{h}_\eta = \frac{1}{2}\left(h_{\eta+1} + h_\eta\right)$. The width $B\left(x^1\right)$ of the electrodes, see Fig. 4, will be used for the further design. Again, the indices of the material parameters are suppressed.

Analogously to the actuator layers one can combine the charge measured by the sensor layers in a symmetric and a skew symmetric way. After some short calculations one obtains

$$Q_\eta^S = -\int_0^L \partial_1 \Gamma_\eta^S \left(x^1\right) \bar{u}^1 \mathrm{d}x^1 + \left\{\Gamma_\eta^S \left(x^1\right) \bar{u}^1\right\}_{x^1\in\{0,L\}} \tag{41}$$

in the symmetric case with

$$\Gamma_\eta^S \left(x^1\right) = \frac{2a_3^{11}}{d_{33}} B_\eta^S \left(x^1\right) \tag{42}$$

and

$$Q_\eta^A = -\int_0^L \left(\partial_1\right)^2 \Gamma_\eta^A \left(x^1\right) \bar{u}^3 \mathrm{d}x^1 + \left\{\partial_1 \Gamma_\eta^A \left(x^1\right) \bar{u}^3 - \Gamma_\eta^A \left(x^1\right) \bar{u}_{01}^3\right\}_{x^1\in\{0,L\}} \tag{43}$$

with

$$\Gamma_\eta^A \left(x^1\right) = \frac{2a_3^{11}\bar{h}}{d_{33}} B_\eta^A \left(x^1\right) \tag{44}$$

in the skew symmetric case. Comparing (41) - (44) with (32) and taking into account (39), we see that the choice

$$\Gamma_\eta^S \left(x^1\right) = \Lambda_\eta^S \left(x^1\right) \,, \qquad \Gamma_\eta^A \left(x^1\right) = \Lambda_\eta^A \left(x^1\right) \tag{45}$$

leads to the *collocated pairing* $\left(U^{\eta}_{S}, \mathrm{d}_0 Q^{S}_{\eta}\right)$, $\left(U^{\eta}_{A}, \mathrm{d}_0 Q^{A}_{\eta}\right)$ of sensors and actuators, which is independent of the kinematic boundary conditions. Thus, we succeeded in fulfilling the demands on the design of a controller as presented in Section 2. Furthermore, we point out that the collocation is independent of the special choice of $\Lambda^{S}_{\eta}\left(x^1\right)$, $\Lambda^{A}_{\eta}\left(x^1\right)$, the only restriction is that (45) is met. This additional degree of freedom can be used for the controller design, see also, (7), (8), (16).

5 The Piezoelectric Cantilever Beam

Let us consider the vertical vibrations of a piezoelectric composite beam, which is fixed at the boundary $x^1 = 0$ and free at the boundary $x^1 = L$. The corresponding kinematic boundary conditions are

$$\bar{u}^3 = 0\ , \quad \bar{u}^3_{01} = 0 \quad \text{at} \quad x^1 = 0\ . \tag{46}$$

We use only two symmetric actuator pairs with skew symmetric power supply, see also Subsection 3.6. One pair has a constant spatial distribution of the electrodes $B^A\left(x^1\right) = B$ such that $\Lambda^A_1\left(x^1\right) = K_1$ with $K_1 > 0$ is met and we take a triangular distribution $B^A\left(x^1\right) = \left(1 - x^1/L\right) B$ for the second pair such that

$$\Lambda^A_2\left(x^1\right) = K_2\left(1 - x^1/L\right)\ , \quad K_2 > 0 \tag{47}$$

is fulfilled. The equations of motion in Hamiltonian form follows as

$$\begin{aligned} \bar{q}^3_{10} &= \bar{p}^3/\mu \\ \bar{p}^3_{10} &= -\Lambda_2 \bar{u}^3_{04}\ , \end{aligned} \tag{48}$$

together with the dynamic boundary conditions

$$\left\{\Lambda_2 \bar{u}^2_{02}\right\}_{x^1=L} + K U^1_A = 0 \tag{49}$$

and

$$\left\{\Lambda_2 \bar{u}^3_{03}\right\}_{x_1=L} - \frac{1}{L} K_2 U^2_A = 0\ . \tag{50}$$

After a short calculation we get

$$\mathrm{d}_0 H_0 = -K_1 U^1_A\ \bar{u}^3_{11}\big|_{x^1=l} - \frac{1}{L} K_2 U^2_A\ \bar{u}^3_{10}\big|_{x^1=L} \tag{51}$$

for the change of H_0 along a trajectory of the system, see also (39).

The collocated sensors follow immediately from (45) such that the electrical charges are related with the displacements in the following manner

$$Q^A_1 = -\left.\bar{u}^3_{01}\right|_{x^1=L}\ , \qquad Q^A_2 = -\left.\bar{u}^3\right|_{x^1=L}\ . \tag{52}$$

Finally, the following fact is worth mentioning. If we rewrite (48) in the following form

$$\frac{\partial}{\partial t}\begin{bmatrix} \bar{u}^3 \\ \bar{p}^3 \end{bmatrix} = \underbrace{\begin{bmatrix} 0 & \frac{1}{\mu} \\ -\Lambda_2\left(\partial_1\right)^4 & 0 \end{bmatrix}}_{\mathsf{A}} \begin{bmatrix} \bar{u}^3 \\ \bar{p}^3 \end{bmatrix} \tag{53}$$

with the differential operator A, then one can prove (see Theorem 6.2.2 in (9)) that the control law, see (7),

$$U_\eta^A = -g_\eta \mathrm{d}_0 Q_\eta^A \ , \quad g_\eta > 0 \ , \quad \eta = 1, 2 \tag{54}$$

stabilizes the piezoelectric beam exponentially.

6 Final Remarks

This contribution shows the application of passivity based control to piezoelectric composite structures. A piezoelectric beam has been investigated in detail within the framework of the Euler-Bernoulli hypothesis. Furthermore, it is shown how one has to extend the passivity based control approach from the finite- to the infinite-dimensional case. Unfortunately, the necessary and sufficient conditions for the stability of the closed loop of finite-dimensional systems are only necessary in the infinite-dimensional scenario. However, for some special cases, like for the cantilever beam under consideration, it is possible to prove exponential stability of the closed loop.

Finally, this contribution is intended to show the benefit of a synergetic design of the intelligent structure together with the control law. We would like to mention that the theory being presented is applicable even if the mathematical model takes additional effects like temperature or nonlinear material behavior into account.

Acknowledgement

This work has been partially done in the context of the European sponsored project GeoPlex with the reference code IST-2001-34166. Additional information is available at http://www.geoplex.cc. Furthermore, it was partially supported by the LCM center at the Johannes Kepler University as a part of the strategic project 4.4.

References

[1] *Abraham, R., Marsden, J.E.* and *Ratiu, T.*: Manifolds, Tensor Analysis and Applications, 2nd edn., Springer, New York, (1988).

[2] *Hebey, E.:* Nonlinear Analysis on Manifolds: Sobolev Spaces and Inequalities, American Mathematical Society, Courant Institute of Mathematical Sciences, New York, (1999).

[3] *Irschik, H., Hagenauer, K.* and *Ziegler, F.*: An Exact Solution for Structural Shape Control by Piezoelectric Actuation, In: Gabbert U. (Ed.), Smart Mechanical Systems-Adaptronics, Fortschrittberichte VDI, Reihe 11, Nr. 244, VDI-Verlag, Düsseldorf, pp. 93-98, (1997).

[4] *Joshi, S.M.*: Control of Large Flexible Structures, Lecture Notes in Control and Information Sciences 131, Springer, London, (1988).

[5] *Khalil, H.K.*: Nonlinear Systems, Macmillan Publishing Company, New York, (1992).

[6] *Klein, W.*: Vierpoltheorie, Wissenschaftsverlag B.I., Leipzig, (1972).

[7] *Kugi, A.*: Non-linear Control Based on Physical Models, Lecture Notes in Control and Information Sciences 260, Springer, London, (2001).

[8] *Kugi, A.* and *Schlacher, K.*: Control of Piezoelectric Smart Structures, In: Preprints of the 3rd-Workshop "Advances in Automotive Control", Karlsruhe, Germany, March 28-30, 2001, Vol.1, pp. 215-220, (2001).

[9] *Liu, Z.* and *Zheng, S.*: Semigroups associated with dissipative systems, Chapman & Hall/CRC Press, Boca Raton, London, (1999).

[10] *Lozano, R., Brogliato, B., Egeland, O.* and *Maschke, B.*: Dissipative Systems Analysis and Control, Springer, London, (2000).

[11] *Nijmeijer, H.* and *van der Schaft, A.J.*: Nonlinear Dynamical Control Systems, Springer, New York, (1991).

[12] *Nowacki, W.*: Dynamic Problems of Thermoelasticity, Noordhoff International Publishing, PWN-Polish Scientific Publishers, Warszawa, (1975).

[13] *Olver P.J.*: Applications of Lie Groups to Differential Equations, Springer, New York, (1993).

[14] *Ortega, R., Loria, A., Nicklasson, P.J.* and *Sira Ramírez, H.*: Passivity-based Control of Euler-Lagrange Systems, Springer, London, (1998).

[15] *Pazy, A.*: Semigroups of Linear Operators and Applications to Partial Differential Equations, Springer, London, (1983).

[16] *Schlacher, K.* and *Kugi, A.*: Control of Mechanical Structures by Piezoelectric Actuators and Sensors, In: Aeyels, D., Lamnabhi-Lagarrique, F., van der Schaft, A. (Eds.) Stability and Stabilization of Nonlinear Systems, Lecture Notes in Control and Information Sciences 246, Springer, London, pp. 275-291, (1999).

[17] *Schlacher, K.* and *Kugi, A.*: Control of Elastic Systems, A Hamiltonian Approach, In: Leonard, N.E., Ortega, R. (Eds.) Preprints of the Workshop on Lagrangian and Hamiltonian Methods for Nonlinear Control, Princeton, New Jersey USA, March 16-18, pp. 80-85, (2000).

[18] *Sepulchre, R., Janković, M.* and *Kokotović P.*: Constructive Nonlinear Control, Springer, London, (1997).

[19] *Tauchert, T.R.*: Piezothermoelastic Behavior of a Laminated Plate, Journal of Thermal Stresses 15, pp. 25-37, (1992).

[20] *Tzou, H.S.* and *Bergman, L.A.*, Dynamics and Control of Distributed Systems, Cambridge University Press, Cambridge, (1998).

[21] *van der Schaft, A.J.*: L_2-Gain and Passivity Techniques in Nonlinear Control, Springer, London, (2000).

Working Group Reports

Report Of Working Group Number One

Submitted by L.A. Bergman[1] and A. DeStefano[2]

[1] *Aeronautical and Astronautical Eng., University of Illinois, 104 South Wright St., 321 E Talbot Cab, MC-236 Urbana, IL 61801, USA*
[2] *Dip. Di Ing. Strutturale, Politecnico di Torino, C.so Duca degli Abruzzi 24, 10129 Torino, Italy*

The Working Group believes the following topics to be of greatest common interest:

1. Sensors
   - Technology driven by guidance, control and navigation applications
   - Dense sensor arrays for locating and isolating damage
   - Distributed sensing for shape control
2. Systems
   - Major systems are aircraft and spacecraft, especially UAVs and smart ordnance
   - Wireless and networkable sensors
   - Fault tolerant sensor networks
3. Analysis
   - Signal processing and analysis software for health and condition monitoring
   - Damage detection
   - Performance evaluation
4. Decision making based on monitored data
   - On-line data mining and analysis

The following are three possible joint demonstration projects

1. In-orbit monitoring of a complex space structure (updated version of NASA's Galileo challenge problem of the 19880s)
2. Condition monitoring of a rotating system and, particularly, of a bladed disk assembly
3. Technology transfer of structural testing and monitoring strategies from aerospace and mechanical to civil engineering applications via an instrumented bridge project

In order to begin to address these topics, it is recommended that the following initial steps be taken:

1. Form an ad hoc Organizing Committee to begin the process
2. Seek relationships with US and European industry and government agencies
3. Solicit concepts for joint problem(s) and experiment(s), and convene a workshop to assess these ideas
4. Establish a first generation joint demonstration project

Report of Working Group Number Two

Submitted by W. D. Iwan[1] and H. Irschik[2]

[1] *Civil Eng., California Institute of Technology, 1200 E. California Blvd,, MC 104-44, Pasadena, CA 91125, USA*
[2] *Division of Technical Mechanics, Johannes Kepler University, Altenbergerstr 69, 4040 Linz, Austria*

The Working Group believes the following topics to be of greatest common interest:

1. Sensors
   - What to measure and how it should be measured
   - Accuracy
   - Reliability
   - Cost
2. Systems
   - Design
   - Connectivity
   - Communication
   - Data management
3. Analysis
   - Health and condition monitoring
   - Damage detection
   - Performance evaluation
4. Decision making based on monitored data
   - Including normal natural events, extreme natural events, and man-made events

The following three components are recommended as a strategy for addressing these topics

1. Analytical and laboratory benchmark problem(s)
2. Full-scale field test structure(s)
3. Supporting infrastructure organization

In order to begin to address these topics, it is recommended that the following initial steps be taken:

1. Form an ad hoc Organizing Committee to begin the process

2. Organizing Committee charged to develop objectives and a strategy for a joint program of research and interaction

3. Convene a joint European-US workshop to design the joint problem(s) and experiment(s)

   Preparation phase:

   - Identify and describe candidate problems and structures
   - Draft possible configurations for funding and management

   Execution phase:

   - Agree on benchmark problem(s) and test structure(s)
   - Design the experiments
   - Identify target funding sources
   - Establish responsibilities and assignments

4. Follow up

   Publish the Workshop Plan

   Establish a permanent Planning/Oversight Committee

   Seek funding to implement the Plan

Conduct scientific/engineering conferences and seminars to highlight technology and results (proceedings to be refereed and published)

Workshop Resolution

The US-European Worshop on Smart Structures and Sensor Technology was successfully held on April 12–13, 2002 in Como and Somo Lombardo Varese, Italy. A total of 46 participants from the US and Europe attended lectures presented during the two-day meeting. The participants of the workshop unanimously resolved that:

- Great opportunities exist for synergysitc cooperative US-European research in smart structures and sensor technology.
- A structured, definitive cooperative mechanism should be quickly established to support such research.
- A joint NSF/ESF scientific committee should be establshed to provide advice, guidance, and planning to the funding agencies. The initial membership of this committee should be as follows:

  – Willfred D. Iwan (Caltech, USA)
  – Billie F. Spencer, Jr. (University of Notre Dame, USA)
  – Talun Yang (ENSCO, Inc., USA)
  – Lucia Faravelli (University of Pavia, European)
  – George Magonette (ELSA Laboratory, European)
  – Hans Irschik (University of Linz, European)

- This planning committee should develop a detailed five-year research plan and report their findings within a 12 months period.
- Research priorities in the areas of sensors, systems, analysis, decision making based on monitored data.
- The following three components are recommended as a strategy for addressing these topics
  1. Analytical and laboratory benchmark problem(s)
  2. Full-scale field test structure(s)
  3. Supporting infrastructure organization

  More detailed information can be found in the attached working group reports.
- Both NSF and ESF are encouraged to commit the necessary funds to facilitate the above referenced planning efforts and the subsequently developed research agenda.
- Significant efforts should be devoted to developing strong partnerships with industry.
- Many emerging technologies, need to prepare the next generation of engineers to take advantage of these opportunities.

AUTHOR INDEX

Abdel-Mooty, M.A.N., *V.3:* 679
Abe, M., *V.1:* 125, *V.2:* 23
Abichou, A., *V.3:* 881
Abou-Kandil, H., *V.3:* 317
Adachi, K., *V.3:* 263
Adachi, Y., *V.3:* 197
Addessi, D., *V.2:* 459
Aghalovyan, L.A., *V.2:* 759
Agrawal, A.K., *V.2:* 863
Aguirre, M., *V.2:* 509
Ahmadi, H., *V.2:* 1089
Aizawa, S., *V.2:* 343
Aizawa, S., *V.3:* 165, 323
Akimoto, M., *V.2:* 269
Al-Dawod, M., *V.3:* 973
Amini, F., *V.3:* 101
André, N, *V.2:* 1127
Anh, N.D., *V.3:* 521
Ansari, F., *V.2:* 895, 917, *V.4:* 45
Antimovski, A., *V.2:* 515
Antretter, T., *V.2:* 561
Argoul, P., *V.2:* 1039
Arikabe, T., *V.3:* 323
Arima, F., *V.3:* 945
Armitage, T., *V.3:* 351
Au, S-K., *V.2:* 1065
Aupérin, M., *V.1:* 205
Azuhata, T., *V.2:* 827
Bahlous, S., *V.2:* 941
Bairrao, R., *V.2:* 1013
Bakule, L., *V.2:* 869
Balandin. D.V., *V.2:* 753
Baratta, A., *V.3:* 873, 887
Barner, A., *V.2:* 903
Barroso, L.R., *V.3:* 495, 659
Bascetta, L., *V.2:* 599
Basedau, F, *V.2:* 903
Battaini, M., *V.2:* 51
Beck, J.L., *V.2:* 1065
Becker, J., *V.4:* 9
Belyaev, S.P., *V.2:* 779
Beni, F., *V.3:* 139
Beomonte, Z., *V.2:* 391
Bergamini, A., *V.2:* 1057
Bergamo, G., *V.2:* 665, 1089
Bergman, L.A., *V.2:* 849, 857, 875, *V.4:* 73, 179
Bergmeister, K., *V.2:* 675
Bernal, D., *V.1:* 265, *V.2:* 1033
Bernelli-Zazzera F.B., *V.2:* 583, 585
Berra, F., *V.2:* 637
Bertocchi, A., *V.3:* 607, 617
Betti, R., *V.2:* 923
Bettinali, F., *V.2:* 1089, 1095
Biella, G., *V.2:* 637
Blakeborough, A., *V.2:* 833
Bonetti, E., *V.2:* 169
Bonfiglioli, B., *V.2:* 911
Boonyapinyo, V., *V.3:*35
Boubaker, M.L., *V.1:* 95
Bourquin, F., *V.3:* 119
Bouvet, C., *V.2:* 381
Bozzo, L., *V.3:* 443
Brailovski, V., *V.2:* 369
Branchet, B., *V.3:* 119
Breitung, K., *V.2:* 467, 765
Bronnimann, R., *V.2:* 897
Bruant, I., *V.3:* 343
Brunetti, G., *V.2:* 689
Bujar, M., *V.3:* 567
Bursi, O., *V.3:* 593
Cacosso, A., *V.3:* 585
Caetano, E., *V.3:* 411
Caffrey, J., *V.3:* 153
Caffrey, J., *V.4:* 91
Caicedo, J.M., *V.2:* 17, 849
Calçada, R., *V.3:* 411
Calloch, S., *V.2:* 381
Cami, R., *V.3:* 607, 617
Camino, J.F., *V.2:* 431
Cantone, G., *V.2:* 637
Cardone, D., *V.3:* 573, 585
Carlson, D.J., *V.1:* 227
Carrier, E., *V.2:* 667

Carvalho, E.C., *V.2:* 1013
Casciati, F., *V.1:* 1, 5, 7, 277, 303, *V.2:* 51, 631, 643, 467, 765, 1137, *V.3:* 747, 753, 951, *V.4:* 133
Castéra, P., *V.2:* 1119, 1127
Caviglione, L., *V.3:* 809
Çelebi, M., *V.2:* 661
Ceravolo, R., *V.2:* 1039, *V.3:* 593, *V.4:* 83
Chakravorty, D., *V.3:* 687
Chang, C.C., *V.2:* 3, *V.3:* 787
Chaouachi, L., *V.2:* 941
Chase, J.G., *V.3:* 495
Chassiakos, A.G., *V.2:* 235
Chen, B.J., *V.2:* 65, *V.3:* 895
Chen, G., *V.2:* 127
Chen, W.S., *V.2:* 65
Chen, Y., *V.3:* 767
Chen, Y-H., *V.2:* 59
Chen, Z.Q., *V.3:* 393, 777
Cheng, F.Y., *V.2:* 127
Cheng, L-Y., *V.3:* 107
Chiang, T.C., *V.2:* 65, *V.3:* 895
Chiesa, S., *V.2:* 637
Choura, S., *V.3:* 881
Chrysostomou, C.Z., *V.2:* 947
Chu, S-Y., *V.2:* 799, 805
Chung, J., *V.2:* 503
Chung, L.L., *V.2:* 65
Ciampi, V., *V.2:* 459, *V.3:* 501
Claus, R.O., *V.2:* 651
Collet, M., *V.2:* 175, *V.3:* 119
Condó, A.M., *V.2:* 157
Corbi, I., *V.3:* 887
Corsi, R., *V.2:* 637
Crail, S., *V.2:* 903
Crewe, A.J., *V.2:* 811
Cunha, Á., *V.3:* 411
Czarnecki, J., *V.2:* 9
Daminelli, R., *V.2:* 637
De Angelis, M., *V.2:* 923, *V.3:* 501
De Canio, G., *V.3:* 579
de Franco, R., *V.2:* 637
De Iuliis, M., *V.3:* 725
de Jager, B., *V.2:* 425
De la Cruz, S.T., *V.3:* 443
De la Sen, M., *V.2:* 89, *V.3:* 203
De Man, P., *V.3:* 305
de Oliveira, M.C., *V.2:* 431
De Stefano, A., *V.1:* 265, *V.2:* 631,1031, 1039, *V.3:* 593, 747, *V.4:* 83, 179
Del Grosso, A., *V.2:* 673, 689, *V.3:* 809
Demetrious, T., *V.2:* 947
DesRoches, T., *V.2:* 375
Di Gerando, A., *V.2:* 51
Dicleli, M., *V.3:*1
Dohi, H., *V.2:* 291
Dolce, M., *V.3:* 573, 585
Domer, B., *V.2:* 449
Dorka, U.E., *V.1:* 273, *V.3:* 217
Drury, D., *V.2:* 811, 819,
Duc, K.T., *V.3:* 753
Dumoulin , C., *V.1:* 205
Durante, F., *V.2:* 391
Dusi, A., *V.2:* 1095
Dye, T., *V.3:* 377
Dyke, S.J., *V.1:* 279, *V.2:* 17, 841, 849, *V.3:* 159
Egawa, K., *V.3:* 673
Eisenberg, J., *V.2:* 473
El-Attar, A., *V.2:* 929, 931
El-Borgi, S., *V.2:* 941, *V.3:* 451, 477, 881
Elgamal, A.W., *V.2:* 1071
Enomoto, T., *V.2:* 269
Erlicher, S., *V.3:* 593
Esteva, L., *V.1:* 33
Faravelli, L., *V.2:* 163, 605, *V.3:* 521, *V.4:* 25
Farrar, C. R., *V.2:* 9, 35
Federici, L., *V.2:* 793
Fedukov, A.A., *V.2:* 753
Feng, D., *V.3:*29
Ferliga, C., *V.2:* 637
Fest, E., *V.2:* 449
Fischer, F.D., *V.2:* 561
Fischer, O., *V.2:* 1113
Foltête, E., *V.2:* 175
Formosa, F., *V.3:* 317
Forni, M., *V.1:* 159, *V.2:* 457, 459, 1095
Fossati, D., *V.2:* 637
Franchioni, G., *V.3:* 665
François, A., *V.3:* 305
Franzolin, R., *V.2:* 1137
Fraser, M., *V.2:* 1071
Fremond, M., *V.2:* 169

Frontini, M., *V.2:* 613
Ftima, M.B., *V.3:* 451
Fujimoto, S., *V.3:* 821
Fujino, Y., *V.1:* 125, *V.2:* 23
Fujita, S., *V.2:* 249, 251, 261, 275
Fujita, T., *V.3:* 323
Fujitani, H., *V.2:* 137, 355, 827
Fukuda, A., *V.2:* 827
Fukui, D., *V.3:* 833
Fukushima, I., *V.2:* 143
Fukuta, T., *V.2:* 185, 1025
Fukuyama, H., *V.2:* 827
Fuller, K., *V.2:* 1089
Furuya, O., *V.2:* 275
Gabbert, U., *V.3:*85
Galimard, P., *V.2:* 1127
Galli, E., *V.2:* 585
Gallone, S., *V.2:* 593
Gang, W., *V.3:* 793
Gao, Y., *V.4:* 3
Garcia, D.L., *V.3:* 431
Gatto, G., *V.2:* 923
Gattulli, V., *V.2:* 95, *V.3:* 337
Gausmann, R., *V.2:* 523
Gavin, H.P, *V.2:* 113, 309, 349
Geier, B., *V.2:* 889
Geier, R., *V.3:* 549
Gerosa, D., *V.2:* 637
Gevorgyan, R.S., *V.2:* 759
Ghrairi, A., *V.2:* 941
Ghulghazaryan, L.G., *V.2:* 759
Giacosa, L., *V.3:* 159
Giorgio, F., *V.2:* 997
Giulia, B., *V.2:* 997, 1103
Giuriani, E., *V.3:* 599
Glaser, S., *V.4:* 19
Glišić, B., *V.2:* 729
Gobetti, A., *V.3:* 311
Gordaninejad, F., *V.2:* 121
Greimann, L., *V.3:* 383
Gubana, A., *V.3:* 599
Guerreiro, L., *V.3:*17
Habel, W., *V.2:* 713, 903
Hague, S., *V.2:* 849
Hahm, D., *V.3:* 457
Hales, M.W., *V.3:* 377
Halling, W., *V.3:* 377
Hamilton, C., *V.2:* 509
Harnpornchai, N., *V.3:* 535
Hata, K., *V.2:* 137
Hatayama, T., *V.3:* 311
Hayakawa, A., *V.3:* 711
He, W.L., *V.2:* 857
Helgeson, R.J., *V.3:* 489
Helmerich, R., *V.2:* 713
Hemez, F., *V.2:* 9
Hera, A., *V.2:* 1121
Hernandez, M.R., *V.3:* 147
Higashino, M., *V.2:* 343
Hirai, J., *V.3:* 179
Hirai, M., *V.3:* 815
Hirata, H., *V.2:* 261
Hisano, K., *V.2:* 291
Hitchcock, G. H., *V.2:* 121
Hiwatashi, T., *V.2:* 137, 355, 827
Hochrainer, M.J., *V.3:*63
Hofmann, D., *V.2:* 903
Holnicki-Szulc, J., *V.3:*69
Honda, M., *V.3:* 179
Hong, H., *V.3:*11
Hong, V.M., *V.2:* 23
Horton, M.A., *V.4:* 19
Hoshino, A., *V.1:* 249
Hou, Z., *V.2:* 1121
Hsu, C-C., *V.2:* 219
Hu, X., *V.3:* 529
Huang, R., *V.4:* 103
Hunt, S., *V.3:* 495
Hwang, J-S., *V.3:* 463
Iemura, H., *V.1:* 85, 175, *V.2:* 45, *V.3:* 191, 197
Igarashi, A., *V.1:* 175, *V.2:* 45
Iiba, M., *V.2:* 137, 185, 355, 827
Ikeda, Y., *V.2:* 115, *V.3:* 127
Ikenaga, M., *V.2:* 285
Inai, E., *V.3:* 469
Inaudi, D., *V.2:* 695, 713, 721, 729, *V.4:* 51
Indirli, M., *V.3:* 607, 617
Inochkina, I.V., *V.2:* 779
Inoue, H., *V.3:* 735
Inoue, K., *V.3:* 179
Inoue, T., *V.3:* 921
Irschik, H., *V.2:* 521, 529, *V.4:* 139, 181
Isalgue, A., *V.2:* 151, 157, 363
Ishiharaguchi, K., *V.2:* 225

Ito, Y., *V.3:* 469, 561
Iwan, W. D., *V.1:* 57, *V.4:* 33, 181
Jabbari, F., *V.2:* 101
Jemai, B., *V.3:* 477
Jerónimo, E., *V.3:*17
Jinping, O., *V.3:* 793
Joghataie, A., *V.3:*45
Johnson, E.A., *V.1:* 259, *V.2:* 349
Jones, A.J., *V.2:* 309
Juhás, M., *V.3:* 665
Juhásová, E., *V.3:* 665
Jung, H-J., *V.2:* 843
Jurukovski, D., *V.2:* 81, *V.3:* 859
Kang, K-S., *V.3:* 463
Kang, T-W., *V.2:* 199
Kanoun, F., *V.2:* 941
Karmakar, A., *V.3:* 687
Kasahara, Y., *V.2:* 261
Kashiwazaki, A., *V.2:* 269
Katayama, T., *V.2:* 1019
Kato, H., *V.2:* 1025
Kato, T., *V.3:* 561
Kawano, H., *V.3:* 711
Kawashima, K., *V.3:*23
Kawatani, M., *V.3:* 933
Kengo, T., *V.3:*91
Kenny, T., *V.2:* 667
Khalil, A.H., *V.3:* 383
Kikuchi, T., *V.2:* 261
Kim, B.H., *V.2:* 503
Kim, J.D., *V.2:* 193
Kimoto, K., *V.3:* 561
Kimura, H., *V.3:* 821, 827
Kinoshita, T., *V.3:* 263
Kiremidjian, A.S., *V.2:* 667
Kitagawa, Y., *V.2:* 185, *V.3:*79
Knapp, J., *V.2:* 713
Knecht, A., *V.2:* 705
Ko, C.H., *V.2:* 59
Ko, J.M., *V.3:* 393, 767, 777
Kobori, T., *V.1:* 9, *V.3:* 915
Kogan, M.M., *V.2:* 753
Koh, H-M., *V.1:* 69, *V.3:* 457, 921
Kohlhoff, H., *V.2:* 713
Kokalevski, M., *V.2:* 515
Kondou, M., *V.2:* 299
Konstantinidis, D., *V.3:* 865
Koo, K.Y., *V.2:* 193
Köppe, H., *V.3:*85
Kosaka, K., *V.2:* 225
Koshida, H., *V.2:* 75
Kosmatopoulos, E.B., *V.2:* 235
Kousaka, R., *V.3:* 561
Kovaleva, A., *V.2:* 745, 747
Krommer, M., *V.4:* 139
Kronenberg, P., *V.2:* 721
Kugi, A., *V.4:* 163
Kullaa, J., *V.3:* 133
Kume, A., *V.2:* 115, *V.2:* 337
Kurabayashi, H., *V.2:* 251
Kurino, H., *V.1:* 211, *V.2:* 75, 143
Kwon, S-J., *V.2:* 199
Lacarbonara, W., *V.3:* 421
Laffi, R., *V.2:* 637
Lagomarsino, S., *V.3:* 139
Lanata, F., *V.2:* 689
Law, K.H., *V.2:* 667
Lazzari, A., *V.2:* 631
Lee, H.S., *V.2:* 207
Lee, I-W., *V.2:* 843
Lee, J.J., *V.2:* 193
Lee, J.W., *V.2:* 193
Lee, S., *V.2:* 243
Lee, S-H., *V.3:* 463
Leitmann, G., *V.2:* 321
Leon, R.T., *V.2:* 375
Lexcellent, C., *V.1:* 95, *V.2:* 175, 381
Li, H-N., *V.2:* 491
Li, J., *V.3:* 973
Liang, Y., *V.2:* 917
Lim, S.W., *V.4:* 149
Limongelli, M.P., *V.3:* 245, 853
Lin, G-L., *V.3:* 507
Lin, S., *V.2:* 101, *V.4:* 39
Lindner, E. *V.2:* 903
Lloyd, G.M., *V.4:* 109
Locatelli, A., *V.2:* 599
Loh, C-H., *V.1:* 115, *V.2:* 219
Loix, N., *V.3:* 305
López, F.A., *V.3:* 443
Lou, M., *V.2:* 127
Lovey, F.C., *V.2:* 151, 157, 363
Lu, L-Y., *V.3:* 507
Lukkunaprasit, P., *V.3:*35
Luo, N., *V.2:* 89, *V.3:* 203, 209
Lus, H., *V.2:* 923

Lutes, L.D., *V.3:* 659
Lynch, J.P., *V.2:* 667
Maenaka, K., *V.3:* 921
Magara, H., *V.3:* 165
Magonette, G., *V.2:* 791, 981, 1007, *V.3:* 139, 627, *V.4:* 65
Maier, A., *V.2:* 539
Makris, N., *V.3:* 865
Manetti, L., *V.2:* 705
Mangerig, I., *V.3:* 171
Manhartsgruber, B., *V.2:* 555
Mañosa, V., *V.3:* 185
Marano, G.C., *V.3:* 245
Marazzi, F., *V.2:* 1007, *V.3:* 139, 627
Marcellini, A., *V.2:* 629, 637
Marioni, A., *V.1:* 159, *V.2:* 51, 459
Martinez, F., *V.3:* 361
Martinez-Rueda, J.E., *V.3:* 223
Masaki, N., *V.2:* 261
Masanori, I., *V.3:* 567
Mascelloni, S., *V.2:* 793
Masic, M., *V.2:* 439
Masoud Z.M., *V.1:* 143
Masri, S.F., *V.2:* 191, 235, *V.3:* 153, *V.4:* 91
Matsumoto, Y., *V.3:* 253
Matsushita, T., *V.2:* 291
Matsuura, T., *V.3:* 469
Matta, E., *V.3:* 747
Mattei, M., *V.3:* 283
Mauriello, D., *V.3:* 649
Mauriello, D., *V.3:*51
McCormack, J., *V.3:* 451
McDaniel, *V.3:* 489
McFarland, D.M., *V.4:* 73
Mechkour, H., *V.3:* 329
Medeot, R., *V.1:* 217, *V.2:* 1077, 1079
Mehr, K., *V.2:* 889
Mekanannapha, C., *V.3:* 403
Melkumyan, M., *V.2:* 771
Mezzi, M., *V.2:* 793
Miara, B., *V.3:* 329
Michaud, V., *V.2:* 399
Midorikawa, M., *V.2:* 827, *V.3:*79
Mimmi, G., *V.2:* 613, 623
Min, K-W., *V.3:* 463
Minowa, C., *V.2:* 827
Mita, A., *V.2:* 1, 29
Mitsumasa, M., *V.3:* 567
Miyazaki, K., *V.3:* 165
Modena, C., *V.3:*11
Molina, F.J., *V.2:* 793, 1007
Monroy, C., *V.3:* 185
Moon, S-J., *V.2:* 849, 875
Mori, F., *V.2:* 75, 143
Morishian, S., *V.3:* 761
Morishita, K., *V.3:* 179
Morita, K., *V.2:* 827
Morrone, A., *V.2:* 637
Motavalli, M., *V.2:* 989
Motosaka, M., *V.2:* 243, *V.3:* 697
Mulas, M.G., *V.3:* 223
Munakata, T., *V.3:* 815, 827
Murai, N., *V.3:* 323
Murano, K., *V.3:* 845
Nader, M., *V.4:* 139
Nagae, K., *V.3:* 945
Nagarajaiah, S., *V.2:* 315, 349
Nagashima, K., *V.2:* 285
Nagata, K., *V.3:* 815, *V.3:* 821, 827
Nakagaki, S., *V.3:* 815, 827
Nakaminami, S., *V.3:* 945
Nakano, T., *V.2:* 291, *V.3:* 945
Nakata, N., *V.1:* 175
Nakaya, N., *V.3:* 761
Narasimhan, S., *V.2:* 349
Nascimbene, R., *V.3:* 311
Nasu, T., *V.3:* 915
Nawakijphaitoon, S., *V.3:* 403
Nawrotzki, P., *V.3:* 229
Nayfeh A.H, *V.1:* 143, *V.3:* 881
Neild, S.A., *V.2:* 811, 819,
Nellen, P.M., *V.2:* 897
Ni, Y.Q., *V.3:* 393, 767, 777:
Nicoletti, M., *V.3:* 573
Nielsen, L.F., *V.2:* 1143
Nigro, D., *V.3:* 573, 585,
Nishimura, H., *V.3:* 275
Nishio, K., *V.3:* 711
Nishitani, A., *V.2:* 115, 307, 337
Nitta, Y., *V.2:* 115, 337
Niwa, N., *V.3:* 635
Oberaigner, E.R., *V.2:* 561
Occhiuzzi, A., *V.2:* 1113, *V.3:* 641, 719
Ocel, J., *V.2:* 375
Ogawa, N., *V.2:* 1019

Ohtani, K., *V.2:* 1019
Ohtori, Y., *V.3:* 297
Ohtsu, M., *V.2:* 225
Okina, Y., *V.1:* 249
Okugawa, M., *V.3:* 673
Olariu, F., *V.2:* 497
Olariu, I., *V.2:* 497
Ornthammarath, T., *V.3:* 535
Osada, A., *V.3:* 827
Osamu, T., *V.2:* 299
Osman, A., *V.2:* 971
Otsuki, M., *V.3:* 821
Pacchiarotti, A., *V.2:* 793
Pagani, M., *V.2:* 637
Palazzo, B., *V.3:* 51, 649, 725
Pan, S., *V.4:* 39
Paolacci, F., *V.3:* 801
Pardi, L., *V.2:* 739, 883
Parducci, A., *V.2:* 793
Park, H.W., *V.2:* 207
Park, K-S., *V.3:* 457, 921
Park, W., *V.3:* 921
Pascale, G., *V.2:* 911
Paulet-Crainiceanu, F., *V.2:* 869
Pavic, A., *V.3:* 351
Peeters, B., *V.1:* 237
Peng, S-X., *V.3:* 927
Pennacchi, P., *V.2:* 613, 623
Perini, R., *V.2:* 51
Perri, J.F., *V.2:* 1039
Petti, L., *V.3:* 649, 725
Petty, T.S., *V.3:* 377
Phanichtraiphop, P., *V.3:*35
Phulé, P.P., *V.2:* 309
Pichler, U., *V.2:* 529, *V.4:* 139
Pieracci, A., *V.2:* 689
Pittas, M., *V.2:* 947
Pizzimenti, A.D., *V.3:* 283
Podestà, S., *V.2:* 643, *V.3:* 139
Poggianti, A., *V.2:* 459
Pong, W., *V.3:* 235
Ponzo, F.C., *V.3:* 573, 585
Poovarodom, N., *V.3:* 403
Pradono, M.H., *V.3:* 191, 197
Preumont, A., *V.3:* 305
Priddy, S., *V.3:* 489
Procaccio, A., *V.3:* 607, 617
Proppe, C., *V.3:* 909
Proslier, L., *V.3:* 343
Pugliese, A., *V.2:* 459
Pujades, L., *V.3:* 443
Qiao, P., *V.3:* 291
Rakicevic, Z., *V.2:* 81, *V.3:* 859
Ranieri, N., *V.3:* 579
Raparelli, T., *V.2:* 391
Rastogi, P.K., *V.2:* 721
Ratier, L., *V.3:* 119
Ravagnani, D., *V.2:* 637
Reed, M., *V.3:* 489
Reichel, D., *V.2:* 903
Reinhorn, A.M., *V.2:* 799, 805
Reithmeier, E., *V.2:* 321
Renda, V., *V.2:* 793, 1007
Renda, V., *V.3:* 627
Renzi, E., *V.3:* 501, 579
Retze, U., *V.3:* 171
Reynier, M., *V.3:* 317
Reynolds, P., *V.3:* 351
Ricciardelli, F., *V.3:* 283, 711
Rigacci, R., *V.2:* 739
Rildova, *V.4:* 97
Riley, M.A., *V.3:* 451, 477
Rivella, D., *V.2:* 1045
Roberts, J.E., *V.1:* 181
Roberts, M., *V.3:* 659
Rocco, P., *V.2:* 599
Rodellar, J., *V.1:* 47, *V.2:* 87, 89, 869, *V.3:* 185, 203
Romeo, F., *V.2:* 95
Rossell, J.M., *V.2:* 869
Rossi, R., *V.2:* 605, *V.3:* 951, *V.4:* 25
Rugginenti, A., *V.3:* 951
Russo, F., *V.3:* 383
Sabia, D., *V.2:* 1045
Sabia, L., *V.2:* 1045
Sade, M., *V.2:* 151, 157
Sadek, F., *V.3:* 451, 477
Sagane, T., *V.3:* 263
Sahakyan A.V., *V.2:* 759
Sahnesaraie, M-A.J., *V.3:*45
Saito, K., *V.2:* 291, *V.3:* 945
Sala, G., *V.2:* 585
Saleh, A., *V.2:* 961
Samali, B., *V.3:* 973
Sanchez-Silva, M., *V.3:* 147
Sang, N.C., *V.3:* 753

Sanli, A., *V.2:* 661
Sano, K-I., *V.3:* 735
Sanò, T., *V.2:* 459
Santa, U., *V.2:* 675
San-Vincente, J-L., *V.3:* 361
Sarbu, D., *V.2:* 497
Sasajima, K., *V.3:* 179
Sato, T., *V.1:* 265, *V.2:* 213, 243
Scattolini, R., *V.2:* 593
Scheidl, R., *V.2:* 555
Schlacher, K., *V.2:* 545, *V.4:* 163
Schmidt, K., *V.3:* 217
Schmitendorf, W.E., *V.2:* 101
Schreiner, U., *V.2:* 903
Schrooten, J., *V.2:* 399
Scivoletto, C., *V.2:* 637
Sedlachek, D.R., *V.3:* 235
Seemann, W., *V.2:* 523, 539
Seiler, C., *V.2:* 1113, *V.3:* 483
Sennhauser, U., *V.2:* 897
Serino, G., *V.2:* 1113, *V.3:* 641, 801
Sethi, V., *V.3:* 291
Seto, K., *V.3:* 253, 263, 297, 833
Settouane, K., *V.2:* 369
Severn, R.T., *V.2:* 983
Shea, K., *V.2:* 449
Shenton, H.W., III, *V.3:* 529
Shibata, H., *V.2:* 1019, *V.3:* 967
Shih, M-H., *V.3:* 959
Shimazaki, M., *V.3:* 311
Shimizu, H., *V.2:* 1121
Shimizu, K., *V.1:* 211
Shimoda, I., *V.2:* 285
Shimodaira, S., *V.3:* 275
Shin, S., *V.2:* 199
Shiozaki, Y., *V.2:* 137, 355, 827
Shiraishi, T., *V.3:* 761
Shoji, G., *V.3:*23
Shono, T., *V.3:* 263
Shouareshi, R.A., *V.4:* 149
Shugo, Y., *V.2:* 185
Shum, K.M., *V.3:* 741
Silverri, E., *V.3:* 337
Silvestri, A., *V.2:* 51
Singh, M.P., *V.3:* 437, *V.4:* 97
Sinha, P.K., *V.3:* 687
Siringoringo, D., *V.2:* 23
Sitar, N., *V.4:* 19
Sittner, P., *V.2:* 399
Skelton, R., *V.1:* 289, *V.2:* 405, 407, 425, 431, 439
Smaoui, H., *V.2:* 941
Smirnov, V., *V.2:* 473
Smith, I.F.C., *V.2:* 449
Smyth, A.W., *V.1:* 265, *V.2:* 235
Socha, L., *V.3:* 909
Soda, S., *V.2:* 137, 355
Sodeyama, H., *V.2:* 355
Sohn, H., *V.2:* 9, 35
Song, G., *V.3:* 291
Soong, T.T., *V.1:* 15, *V.2:* 799, 805, *V.3:* 431
Sophocleous, A., *V.2:* 955
Spadoni, B., *V.2:* 459
Spencer, B.F. Jr., *V.2:* 843, *V.3:* 777, *V.4:* 3
Spizzuoco, M., *V.2:* 1113, *V.3:* 483
Stasis, A., *V.2:* 947
Stoten, D.P., *V.2:* 811, 819, 983
Strauss, A., *V.2:* 675
Sugimura, Y., *V.3:* 945
Sugiyama, T., *V.2:* 75, 143
Sun, C., *V.2:* 917
Sun, L., *V.3:* 29
Sun, Z., *V.2:* 3
Sunakoda, K., *V.2:* 137
Suwa, M., *V.2:* 75, 143
Suzuki, D., *V.3:* 711
Suzuki, H., *V.3:* 697
Suzuki, S., *V.2:* 261
Suzuki, U., *V.2:* 1121
Suzuki, Y., *V.2:* 45
Symans, M.D., *V.2:* 331
Syrmakezis, C.A., *V.4:* 155
Syrmakezis, K., *V.2:* 955
Tagami, J., *V.1:* 211, *V.2:* 75
Tajima, H., *V.3:* 263
Takanashi, K., *V.3:* 165
Takeuchi, S., *V.3:* 561
Takhira, S., *V.2:* 29
Tamaki, T., *V.3:* 711
Tanaka, H., *V.3:* 833
Tanaka, K., *V.2:* 561
Tanner, N.A., *V.2:* 35
Tashkov, L., *V.2:* 515
Tateno, T., *V.3:* 697

Tatone, A., *V.3:* 337
Taucer, F., *V.2:* 1007
Taylor, C.A., *V.2:* 983
Tento, A., *V.2:* 637
Terriault, P., *V.2:* 369
Teshigawara, M., *V.2:* 827
Tetsuya, O., *V.3:*91
Thanh, V.D., *V.3:* 521
Tirelli, D., *V.2:* 1007
Tomura, T., *V.2:* 355
Torelli, G., *V.3:* 951
Torra, V., *V.2:* 149, 151, 157, 361, 363
Toyama, K., *V.3:* 165, 323
Toyooka, A., *V.2:* 45
Toyota, K., *V.3:* 945
Trajkov, T.N., *V.3:*85
Tsai, C.S., *V.2:* 65, *V.3:* 895
Tse, T., *V.3:* 787
Tseng, T-C., *V.2:* 219
Tsuyuki, K., *V.2:* 299
Turan, G., *V.2:* 849, 857
Ubaldini, M., *V.2:* 51
Ueda, H., *V.2:* 261
Vakakis, A.F., *V.4:* 73
Valfrè, G., *V.3:* 809
Van der Auweraer, H., *V.1:* 237
Varadarajan, N., *V.2:* 315
Vehí, J., *V.2:* 89, *V.3:* 203, 209
Venini, P., *V.3:* 311, 515
Vestroni, F., *V.3:* 421
Villamizar, R., *V.3:* 209
Villaverde, R., *V.2:* 509
Vitantonio, R., *V.2:* 1051
Volkov, A.E., *V.2:* 779
Voulgaris, P.G., *V.2:* 857, 875
Vurpillot, S., *V.2:* 729
Wada, A., *V.3:* 561
Wagg, D.J., *V.2:* 811, 819,
Wait, J.R., *V.2:* 35
Walha, M.A., *V.3:* 477
Wang, M.L., *V.4:* 109
Wang, S-G., *V.2:* 107
Wang, X., *V.2:* 121
Wang, X.Y., *V.3:* 393
Watanabe, H., *V.3:* 711
Watanabe, T., *V.3:* 297, *V.3:* 833
Wenzel, H., *V.2:* 881
Wenzel, *V.3:* 549
Wiklo, M., *V.3:*69
Willford, M., *V.3:* 351
Williams, M.S., *V.2:* 833
Williamson, D., *V.2:* 407
Winkler, B., *V.2:* 555
Wipf, T.J., *V.3:* 383
Wolfe, R., *V.3:* 153
Wolfe, R.W., *V.4:* 91
Wongprasert, N., *V.2:* 331
Wright, J., *V.3:* 351
Wu, J-C., *V.3:* 107
Wu, T-C., *V.2:* 219
Wu, W-H., *V.3:* 927
Wu, X-X., *V.2:* 491
Xing, L., *V.3:* 437
Xu, Y.L., *V.3:* 741
Yamada, T., *V.1:* 211
Yamaguchi, S., *V.2:* 115, 337
Yamamoto, H., *V.2:* 243
Yamamoto, M., *V.2:* 343
Yanagida, H., *V.2:* 269
Yang, G.L., *V.4:* 119
Yang, J.N., *V.1:* 279, *V.2:* 101, 857, *V.4:* 39
Yang, T-L., *V.4:* 127
Yang, Y.J., *V.2:* 503
Yasuda, K., *V.3:* 711
Yohji, I., *V.3:* 567
Yoneda, M., *V.3:* 367
Yoshida, I., *V.2:* 213
Yoshida, K., *V.3:* 821, 845
Yoshida, O., *V.3:* 159
Yoshizawa, T., *V.3:* 561
Yoshizumi, F., *V.3:* 735
Yuan, S., *V.4:* 103
Yuen, K-V., *V.2:* 1065
Yun, C.B., *V.1:* 265, *V.2:* 193
Zaghw, A.H., *V.2:* 961
Zambrano, A., *V.3:* 703
Zanardo, G., *V.3:*11
Zanon, P., *V.3:* 593
Zehetleitner, K., *V.2:* 545
Zeng, T., *V.2:* 651
Zhang, L., *V.2:* 651
Zhao, L., *V.3:* 529
Zheng, G., *V.3:* 767
Zonta, D., *V.2:* 1071, *V.3:* 593
Zuccaro, G., *V.3:* 541